D1443285

Cracking the Particle Code of the Universe

Cracking the Particle Code
of the Universe

The Hunt for the Higgs Boson

JOHN W. MOFFAT

OXFORD
UNIVERSITY PRESS

OXFORD
UNIVERSITY PRESS

Oxford University Press is a department of the University of Oxford.
It furthers the University's objective of excellence in research, scholarship,
and education by publishing worldwide.

Oxford   New York
Auckland   Cape Town   Dar es Salaam   Hong Kong   Karachi
Kuala Lumpur   Madrid   Melbourne   Mexico City   Nairobi
New Delhi   Shanghai   Taipei   Toronto

With offices in
Argentina   Austria   Brazil   Chile   Czech Republic   France   Greece
Guatemala   Hungary   Italy   Japan   Poland   Portugal   Singapore
South Korea   Switzerland   Thailand   Turkey   Ukraine   Vietnam

Oxford is a registered trademark of Oxford University Press
in the UK and certain other countries.

Published in the United States of America by
Oxford University Press
198 Madison Avenue, New York, NY 10016

Library of Congress Cataloging-in-Publication Data
Moffat, John W.
Cracking the particle code of the universe : the hunt for the Higgs boson / John W. Moffat.
pages cm
Includes bibliographical references.
ISBN 978–0–19–991552–1 (alk. paper)
1. Higgs bosons.   2. Particles (Nuclear physics)   I. Title.   II. Title: Hunt for the Higgs boson.
QC793.5.B62M64 2014
539.7'21—dc23
2013029713

9 8 7 6 5 4 3 2 1
Printed in the United States of America
on acid-free paper

To Patricia, with love

CONTENTS

In April 2008, I traveled to Geneva for a week to visit the new facilities at the European Organization for Nuclear Research, known as CERN—the large hadron collider (LHC), the biggest and most expensive scientific experiment ever built. My wife Patricia and I stayed in Ferney-Voltaire, just across the border from Switzerland. From our hotel window, we could see the jagged white caps of the Jura Mountains rising over the charming little French town. Driving each day to CERN on a narrow country road, we crossed the French–Swiss border twice, and enjoyed the views of meadows, farms, and villages.

This was a pilgrimage that I had intended to make for years. I wanted to see with my own eyes the enormous particle accelerator whose experiments promised to answer important questions and settle long-standing disputes in particle physics. The $9 billion machine would be hunting for several discoveries. One is the so-called *supersymmetric particles*, necessary components of the Holy Grail of physics, the yet-to-be-discovered unified theory of all the forces of nature. Another possible discovery is *extra space dimensions*, beyond the three we inhabit, which are required by string theory. Thirdly, the LHC experimentalists hope to find the elusive "dark matter" particles that the majority of physicists believe make up more than 80 percent of the matter in the universe. The LHC might even succeed in producing *mini black holes* during its proton–proton collisions, a possibility that, stoked by media reports, initially flared up into worldwide hysteria, with certain individuals attempting to close down the LHC by litigation.

But most important, the LHC was built for the main purpose of finding the final puzzle piece required to confirm the standard model of particle physics, the so-called *Higgs boson*. Within the almost half-century-old, widely accepted theory describing the subatomic elementary particles and the three subatomic forces (excluding gravity), the mother of all particles was the Higgs. In this standard theory, on which thousands of physicists had worked and contributed to since the mid 1960s, the Higgs particle, or boson, or field, gave all the other elementary particles their masses back near the very beginning of the universe.[1]

Most physicists believe that elementary particle masses come about because of the special relationship the Higgs boson and its field enjoy with the vacuum. This is the physical state of lowest energy existing at all times, including at the beginning of the universe. The modern concept of the vacuum, quantum mechanics tells us, is not simply a void containing nothing; it is a teeming cauldron of particles and antiparticles flashing into existence and immediately annihilating one another. According to the standard model of particle physics, without the Higgs boson, the basic constituents of matter—the quarks and leptons—would have no mass.[2] Physicists sometimes liken the Higgs field to a river of flowing molasses or a viscous kind of ether permeating space. When the original, massless elementary particles moved through it at the beginning of the universe, they picked up sticky mass. The idea that we require a Higgs field originated with the physics of low-temperature superconductors—mainly with Russian physicists Lev Landau and Vitaly Lazarevich Ginzburg, and later with the American Nobel laureate Philip Anderson. The Higgs is so important to the standard theory of particle physics that it has been nicknamed the *God particle*.

To date, all the other predictions of the standard model have been validated by experiments. During the 1970s, experiments at the Stanford Linear Accelerator (SLAC) verified Murray Gell-Mann's and George Zweig's 1960s-era predictions of quarks. And, in 1983, the so-called *W* and *Z bosons*, the predicted carriers of the weak nuclear force of radioactive decay, were discovered at CERN. Over time, three basic families of quarks and leptons have been detected by colliders, and the carrier of the strong nuclear force, the gluon, was also verified to exist. Today, only the Higgs boson remains to be found. Its detection had to wait until a much larger accelerator could be built to create the incredibly high energies that would be necessary to detect this massive particle, energies equivalent to the temperature at the beginning of the universe, a fraction of a second after the Big Bang.

On July, 4, 2012, two groups associated with the compact muon solenoid (or CMS) and a toroidal LHC apparatus (ATLAS) detectors at the LHC announced the discovery of a new boson at about 125 GeV, that is, at a mass

1. An elementary particle is not composed of other particles bound together by a force; the electron, for example, is an elementary particle, but the proton and neutron are not, being composed of quarks.

2. Leptons are weakly interacting particles such as the electron, muon, and tau, and the neutrinos.

of 125 billion electron volts.[3] This boson appeared to have the properties of the standard-model Higgs boson, but the experimental groups were cautious about identifying it as the Higgs boson. Although the majority of physicists now believe that the new boson is the Higgs boson, we are currently waiting for LHC experimentalists to complete the analysis of the 2012 data and for the accelerator to start up again in 2015 to collect even more data to confirm definitively the identity of the new boson.

The standard theory of particle physics is one of the most successful physics theories of all time. It is on par with James Clerk Maxwell's electromagnetism, Isaac Newton's gravitation, Albert Einstein's general theory of relativity, and the theory of quantum mechanics, which was a cooperative venture by about a dozen physicists during the early 20th century. Even though the final mechanism that keeps the whole edifice together, the Higgs boson, had not yet been detected in 2008 when I visited CERN, the majority of physicists accepted the theory almost without question, and assumed that the discovery of the Higgs would be inevitable, almost a formality. The Nobel committee, too, had already given out five Nobel Prize(s) in Physics to theorists and experimentalists working on the standard model of particle physics, even though the Higgs had not yet been detected and, therefore, the theory had not been fully proved. Finding the Higgs boson was considered such a certainty in 2008 that a dispute had arisen about who would get the Nobel Prize for predicting it. In 1964, during a three-month period, a total of six physicists published short papers in *Physical Review Letters* promoting a way of giving elementary particles their masses. These physicists were François Englert, Robert Brout, Peter Higgs, Carl Hagen, Gerald Guralnik, and Tom Kibble. Because the Stockholm committee can award one prize to no more than three people, if the Higgs was discovered, they would have quite a dilemma deciding among these six physicists. All were eagerly awaiting their Nobel Prizes, and trying to stay alive until the Higgs was found, because the Nobel committee is also constrained by the rule that no prizes can be given posthumously.[4] To complicate matters further, there is a seventh physicist, Philip Anderson at Princeton University, who published a seminal paper in 1963 proposing what is now called the *Higgs mechanism* to give masses to particles. It is worth noting, however, that only the English

3. In Einstein's special relativity, energy is equivalent to mass through his famous equation $E = mc^2$. We express particle masses in units of energy, such as electron volts (eV), thousands of electron volts (KeV), millions of electron volts (MeV), billions of electron volts (GeV), and trillions of electron volts (TeV). For example, the electron has a mass of 0.5 MeV, the proton has a mass of 938 MeV, the W boson has a mass of about 80 GeV, and the top quark has a mass of about 173 GeV.

4. Sadly, Robert Brout died in May 2011, which does, however, make the committee's decision somewhat easier.

physicist Peter Higgs had predicted explicitly the existence of an actual particle in his paper.

But what if the standard theory of particle physics was not correct and the particles derived their masses in some other way? Or had their masses right from the beginning, with no intervention necessary by a God particle? What if the Nobel committee had been premature with its awards for the standard model of particle physics? What if the enticing hints of the Higgs boson at 125 GeV either evaporated or turned out to be another new particle entirely? What if the enormous LHC never found the Higgs boson after all?

This was the second reason for my pilgrimage to Geneva in 2008. Along with my research in gravitation and cosmology, I had been working on an alternative theory of particle physics since 1991, and there was no Higgs boson in my theory. In the mathematics of my *alternative electroweak theory*, all the elementary particles were massless at the beginning; but, except for the massless photon, their masses were then generated not by a single particle with its associated special vacuum features, but by the usual dynamical processes of quantum field theory.[5] That is, the primary observed elementary particles such as the quarks and leptons, and the W and Z bosons conspired—through the quantum field dynamics of self-energies—to produce their own masses. Moreover, my theory did not require the discovery of *any* new particles beyond the already observed ones in the standard model. For example, it did not require any hypothesized particles of supersymmetry, which had, over the years, become a large research industry. My theory seemed to me an economical description of the elementary particles, fields, and forces. I was not the first or only physicist to try to construct an alternative electroweak theory. Attempts to avoid introducing scalar fields and the Higgs mechanism into the standard model had been proposed during the early 1970s by, among others, Roman Jackiw and Kenneth Johnson at the Massachusetts Institute of Technology (MIT).[6]

I called the talk that I gave to the theory division at CERN during that week in April 2008 "Electroweak Model without a Higgs Particle," a provocative title to the theorists and experimentalists who had been working for years, in some cases *decades*, on the standard model of particle physics, on the Higgs mechanism, and on figuring out exactly how the enormous new machine might detect it. Two weeks

5. In modern particle physics, quantum mechanics and relativity are united in quantum field theory. Each elementary particle has an associated field, and this field is quantized so that the particle field is consistent with quantum mechanics and relativity. Quantum field theory is used to do calculations in particle physics.

6. R. Jackiw and K. Johnson, "Dynamical Model of Spontaneously Broken Gauge Symmetries," *Physical Review*, D8, 2386–2398 (1973).

before my talk, the LHC had had its official opening. Present at the launch was Peter Higgs, retired professor of physics at the University of Edinburgh. During an interview at the LHC opening, a journalist asked Higgs how he felt about having $9 billion spent on finding a particle named after him, and did he think they would find it? Peter replied, "I'm 96 percent certain that they will discover the particle."

****

  On the day of my talk, Patricia and I parked our rental car outside the visitors' entrance to CERN and made our way across the sprawling research compound to the theory building. I remembered the building well from my days visiting CERN when I was on sabbatical leave at Cambridge University in 1972 and also when I was a visiting fellow at CERN in 1960/1961. As we mounted the scuffed stairs to the second-floor administrative office, it struck me that the building had changed little in nearly half a century. The halls were dark and dingy, exuding an air of weariness and neglect. I soon discovered that the toilet roll holder in the men's room was broken, just as it had been those many years ago during my last visit. Apparently the $9 billion price tag for the LHC had not included renovations to the theoretical physicists' working quarters.

  My talk was scheduled for two o'clock in the large seminar room on the second floor. At five minutes to two, I stood waiting at the front of the hall with the CERN theoretician who had organized the seminar, staring out at the large amphitheater with its rising rows of desks that was completely empty except for my wife sitting in the fourth row. I thought, *My goodness! Maybe no one is going to turn up because of my audacious title.* Then, at three minutes to two, about 50 theorists and experimentalists swarmed into the room and sat down. Five minutes into my talk, I noticed a professor from New York University (NYU), with whom I had had encounters before, walk in and sit at the back. Sure enough, he soon erupted into loud technical protestations about my theory. A verbal duel ensued and, thanks to many years of experience at such seminars, I was able to subdue him sufficiently to continue. There were other interruptions, and I could feel the audience growing increasingly hostile to the idea of a particle physics theory without a Higgs particle. Near the end of my talk, the NYU professor again got excited and explained, as he understood it, that the Higgs particle existed in my theory, but in an altered way. Alvaro de Rújula, a senior theorist at CERN, who was sitting next to him in the back row, could be overheard saying quite loudly, "Moffat does not *have* a Higgs particle in his theory!" This seemed to flummox our friend and from then on, he kept quiet.

  After my talk, Guido Altarelli, a senior physicist in the theory division who is an expert on the electroweak theory and the standard model of particle physics, declared dramatically, "If Moffat is correct, and no new particles are discovered, then that's the end of particle physics! What are we going to do?"

"You can all retire!" I quipped, trying to break the tension in the room with a joke—but no one laughed.

So I pursued Altarelli's point. "What do you mean, Guido, that this will be the end of particle physics?"

Guido, a strongly built, tall Italian, turned to the audience and said, "What I mean is that governments will no longer give us money to build new accelerators and continue our experiments. In that sense, it's the end of particle physics."

"But this is not the way physics works," I said, genuinely astonished. "Not finding the Higgs and, in fact, not needing to find any other exotic particles would open up many new questions about the nature of matter. New mysteries would unfold. This kind of situation often leads to a revolution in physics."

I reminded the audience of the famous Michelson–Morley experiment in the United States during the late 19th century that failed to detect the "ether," which virtually every scientist of the day accepted as real. To those scientists, it had seemed obvious that a medium was necessary for electromagnetic waves to propagate. The waves had to be moving through *something*. But Michelson and Morley's ingenious experiment turned up nothing where the ether should have been. Although the physics community of the time was shocked, it was certainly not the end of physics. It heralded a new beginning, with Einstein's subsequent discovery of special relativity, and Max Planck's discovery of the quantization of energy, which led to quantum physics. A similar boost to the whole enterprise of physics, especially particle physics, would occur if the LHC did not detect the Higgs boson, I concluded.

But neither Altarelli nor the rest of the audience seemed convinced. The questions and comments at the end of my talk continued to be skeptical in tone, even hostile.

****

I was disappointed by the reception of my alternative electroweak theory at the CERN theory group, but I was not surprised. I knew that even expressing doubts about a prevailing paradigm causes conflict. Physicists are like most other human beings; they become emotionally invested in the truth and beauty of the theoretical structures they have helped to build. To contemplate that those structures might be faulty or incomplete simply pulls the rug out from under their feet. On the other hand, the standard model with the Higgs boson has very attractive features, such as the elegant prediction during the 1960s of the W and Z bosons by Sheldon Glashow, Steven Weinberg, and, independently, by Abdus Salam. The discovery of the W and Z bosons at CERN in 1983 confirmed this remarkable prediction. Moreover, the inclusion of the Higgs boson into the unified electroweak theory guaranteed that one could perform finite calculations of physical quantities in the standard model.

Soon, in a matter of years rather than decades, the answers to the Higgs boson puzzle would be known. The LHC's beams of protons would smash together, spraying subatomic debris into the largest and most sophisticated detectors ever built. Physicists would analyze the statistical patterns of those collisions, an immense amount of data. Eventually, after a great many such experiments, CERN would make its announcement. Either the LHC would have finally discovered the Higgs boson, and therefore proved that the standard model of particle physics was correct in all respects, or the LHC would not have found a Higgs boson, and therefore the standard model would have to be reexamined and changed. If the latter happened, I'd told the skeptical crowd of CERN physicists at the end of my talk, I would be standing at the head of the queue called "Other Ideas," offering my alternative electroweak theory without a Higgs boson for serious consideration.

Later in the week, Patricia and I joined one of the last tours of the new accelerator before it was turned on. Once the machine was operating and the protons were chasing around the 27-km circumference of the LHC, then the radiation level would be prohibitive and access to the machine would be severely restricted. We stood on the viewing deck of the enormous ATLAS detector, feeling overwhelmed by its sheer size. The amount of iron used to build just this one detector was equivalent to all the iron in the Eiffel Tower. Colorful cables wound through the complicated electronic parts of the detector, which appeared to us to be about the size of a European cathedral. Together with three other main detectors, ATLAS was the place where the protons would collide and produce a huge amount of particle debris, which would then be analyzed for years by a worldwide grid of computers.

****

At this writing, five years later, the LHC has been shut down for two years of maintenance and upgrading to an energy of 13 to 14 TeV. During the past two years, there has been tremendous excitement about what the LHC may have discovered, with many physicists already popping the champagne corks to celebrate the discovery of the Higgs boson. The data accumulated before the LHC shutdown have been analyzed and found to be consistent with the standard-model Higgs boson. However, as we will learn in this book, there remain critical experimental issues to determine exactly what the new Higgs-like boson is, which need to be resolved after the machine starts up again in 2015.

Cracking the Particle Code of the Universe

What Is Everything Made Of?

The fifth-century BC Greek philosopher Leucippus, and his pupil Democritus (460 to 370 BC), presaged in broad strokes by more than 2,000 years the standard model of particle physics. They claimed that matter could be broken down to a basic unit. This claim was in contrast to Aristotle (384 to 322 BC), whose ideas, based on Plato's, soon became the establishment view of the day, and who taught that the basic elements of the universe were the earthly elements fire, water, earth, and air, plus the heavenly element, ether. Leucippus and Democritus called their basic unit *atom*, which means "indivisible" or "uncutable" in Greek. By 1964, with the publication of the quark model by Murray Gell-Mann and the independent proposal by George Zweig of *aces*, the ancient Greek idea of a basic unit of matter had finally arrived—and the quarks, along with the leptons, would eventually be heralded as the basic units of all matter.[1]

FINDING THE BUILDING BLOCKS

Particle physics has a long history, reaching back to the 19th century and the discovery of atoms. The reductionist view that matter is constituted of tiny, indivisible units received its first serious validation in 1897, when J. J. Thomson of Cambridge University discovered the electron. Ernest Rutherford's discovery around 1919 of the proton—the positively charged particle at the center of the hydrogen atom, with Thomson's negatively charged electron buzzing around it—introduced another important unit of matter. It was not until 1932, when

---

1. Several authors have questioned the idea that quarks are the final, basic, and indivisible units of matter by claiming that quarks are composed of "prequarks." The large hadron collider has searched for such prequarks, or preons, and the possible composite nature of quarks, and so far has not found evidence for them.

James Chadwick discovered the neutron with the first particle accelerator, that the dominance of the proton and electron as the only basic units of matter was broken. This was the birth of nuclear physics; the nuclei of more complicated atoms and molecules were discovered to be composed of different numbers of bound protons and neutrons. The radioactive beta decay—or transmutation of matter first discovered by Henri Becquerel in 1896—of some of these nuclei released a further unit of matter, called the *neutrino*, which was a massless, electrically neutral particle. At the time of its discovery, it was believed to have zero mass, and, obeying Einstein's special relativity theory, it moved at the speed of light. The neutrino was first proposed by Wolfgang Pauli in 1931 to explain the observed missing energy in radioactive decays. He surmised that an unseen neutral particle was carrying it off. The neutrino was not confirmed experimentally until 1959, by Fred Reines and Clyde Cowan.

The famous Dirac equation, invented by Paul Dirac in 1928, provided a particle-wave description of the electron in accordance with Einstein's special relativity. However, it had an interesting by-product; it also predicted the existence of antimatter. Carl Anderson's detection in 1932 of the "positron," a positively charged electron, was a triumph for Dirac. Eventually, this discovery led to the understanding that all the basic constituents of matter had antimatter partners. For example, the proton's antiparticle is the antiproton.

As accelerators were developed that collided particles at higher energies, more particles and their antimatter partners were discovered. The muon, a kind of heavy electron, was found by Carl Anderson and Seth Neddermeyer at Caltech in 1936. This was followed by the discovery of the mesons, the particles that were, at that time, considered to be the carriers of the strong force holding atomic nuclei together, much as the photon carries the electromagnetic force. In 1935, Japanese physicist Hideki Yukawa had predicted that the protons and neutrons were bound in the atomic nucleus by the exchange of mesons, and after World War II, in 1947, Cecil Powell, César Lattes, and Giuseppe Occhialini of the University of Bristol detected these particles in cosmic rays.

GROUPING THE BUILDING BLOCKS

The electron, proton, neutron, neutrino, antimatter, meson, muon,...! What next? By the mid 1950s, the number of newly discovered particles at accelerators was so large that physicists began referring to them as the *particle zoo*. To make sense of the abundance of particles, physicists began categorizing them, using the ideas from group theory, which had been promoted by Hermann Weyl, John von Neumann, and other mathematicians. Murray Gell-Mann and Yuval Ne'eman played a significant role in this effort. Gell-Mann was a dominant

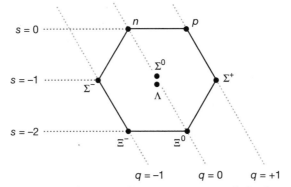

Figure 1.1 Gell-Mann's Eightfold Way of baryons. Here, s stands for the quantum number strangeness, and q stands for electric charge.
SOURCE: tikalon.com.

figure in particle physics during the 1960s and 1970s. He was a professor of physics at Caltech, and won the Nobel Prize in 1969 for his contributions to particle physics and the discovery of quarks.

During the early 1960s, before the advent of the quark model, Gell-Mann developed what he called the *Eightfold Way*, alluding to the Buddhist Noble Eightfold Path of right thought and actions leading to enlightenment. He proposed that what were now called the *baryons* (the proton, neutron, and their cousins with spin ½ and ³⁄₂) fitted into octets (Figure 1.1) and decuplets (Figure 1.2)—namely, patterns of eights and 10s—of the symmetry group from group theory called SU(3).

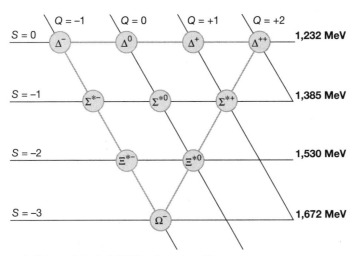

Figure 1.2 Gell-Mann's Eightfold Way decuplet of baryons.
SOURCE: Wikipedia Commons.

One of the baryons in the decuplet was missing. There was nothing in the place where a particle should be at the base of the triangle. Gell-Mann, confident of his patterned scheme, predicted that the missing particle must exist. He called it the Ω^-—the Omega minus, a negatively charged resonance.[2] Gell-Mann's classification scheme was validated in 1964 when a team of physicists from Brookhaven, the University of Rochester, and Syracuse University discovered the Omega minus particle.

Before wading any further into particle physics terminology, some brief explanations are in order. *Spin* is one of several parameters that characterize the different particles. It is an intrinsic quantum number that was discovered during the early development of quantum mechanics by Samuel Goudsmit and George Uhlenbeck in 1925. Spin is a degree of freedom that has meaning only in quantum mechanics. Its classical counterpart is the angular momentum of a body, such as the spinning of a top. However, the quantum mechanical spin does not have a classical interpretation. The spin of a particle is described by being either up or down in direction. Its magnitude can be integer values such as 0, 1, or 2 for bosons, and half-integer values such as ½ and ³⁄₂ for fermions. In physical units, the spins of the particles are given in multiples of Planck's constant h; so, for example, the spin of the electron is ½ h.

Some of the other properties that characterize particles are electric charge, mass, and parity (left- or right-handedness). Group theory will be described in more detail in Chapter 3. Suffice it to say here that, of the several mathematical groups that have been applied successfully to particle physics, the group SU(3) is one of the most important groups used by physicists to categorize particles.

In addition to the baryon octet and decuplet categories, there are octets of mesons having spin 0 (Figure 1.3) and spin 1 (Figure 1.4). Most of the particles that fitted into these meson octets, such as the eta meson and the electrically neutral K mesons, were already known in the 1960s or, if not, were soon discovered. Two of these particles were the rho and omega mesons with spin 1, which were thought, at the time, to be important force-carrying particles that bound protons and neutrons together.

A basic feature of the group SU(3) is the triplet. It implies that there could be three fundamental constituents of all the known hadrons, a category that includes baryons and mesons; *hadros* means "heavy" in Greek. Back in 1956, Japanese physicist Shoichi Sakata proposed that three particles are the building blocks of all others: the proton, neutron, and "Lambda (Λ) particle." Initially,

---

2. At this time, particles were identified by "bumps" appearing in the measured cross-sections of particle collisions in accelerators. These bumps had a mass and a width determined by a formula obtained by physicists Gregory Breit and Eugene Wigner, and were called *resonances* in the cross-sections.

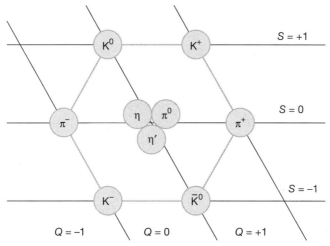

Figure 1.3 Gell-Mann's Eightfold Way spin-0 pseudoscalar (negative-parity) meson nonet (octet plus a singlet).
SOURCE: Wikipedia Commons.

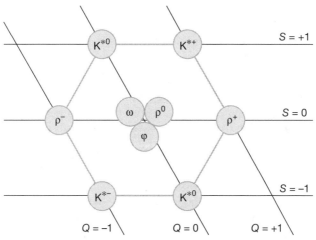

Figure 1.4 Gell-Mann's Eightfold Way spin-1 pseudoscalar meson nonet (octet plus a singlet).
SOURCE: Wikipedia Commons.

Murray Gell-Mann went along with this idea. The proton and neutron, collectively called *nucleons*, were interpreted in terms of the quantum number called *isospin*, which is a property of the proton- and neutron-like spin. Isospin means that hadrons with similar masses and the same spin but different electric charges were identical as far as the strong interactions within the nucleus were concerned.

The Lambda particle is an electrically neutral, short-lived particle discovered in 1947. Sakata included the Lambda in the triplet to account for the quantum number "strangeness." Abraham Pais first proposed a quantum number in 1952 to explain that certain particles, such as the Lambda, even though they were produced copiously in particle interactions, decayed slowly, having lifetimes typical of weak radioactive decays. This quantum number was dubbed *strangeness* by Gell-Mann. The idea of the triplet of proton, neutron, and Lambda was that all the other hadrons in the particle zoo were made up of these three particles bound together. This idea did not fit with the experiments at the time and was abandoned. However, the idea of a stable triplet of basic constituent particles for matter persisted.

"THREE QUARKS FOR MUSTER MARK!"

In 1963/1964, Murray Gell-Mann speculated on the basis of a suggestion by Robert Serber, a professor at Columbia University, that there were three basic constituents of all matter, which had not yet been observed. Gell-Mann understood that because these three constituent particles would make up protons and neutrons, which had a unit of positive electric charge and zero charge respectively, the charge of the constituent particles had to be fractional, ⅔ and ⅓, to create a plus-one or zero charge. Because no one had ever observed such fractional electric charges, this seemed an absurd idea. However, Gell-Mann continued to pursue the idea and published a letter in *Physics Letters* in 1964 titled, "A Schematic of Baryons and Mesons," proposing that these fractionally charged particles played a mathematically important role in the basic constituents of matter.[3] In his letter, he called the particles *quarks*, taken from James Joyce's *Finnegan's Wake*: "Three quarks for Muster Mark!"

Meanwhile, George Zweig, who was a Caltech graduate visiting CERN as a postdoctoral fellow, proposed independently at about the same time an idea that was very similar to Gell-Mann's quark model. He called his basic constituents of the triplet *aces*. However, he did not publish his paper because senior physicists at CERN considered it to be too speculative. Zweig later recalled:

The reaction of the theoretical physics community to the ace [quark] model was generally not benign. Getting the CERN report published in the form that I wanted was so difficult that I finally gave up trying. When the physics department of a leading university was considering an appointment for me, their senior theorist, one of the most respected spokesmen for all of theoretical physics, blocked the appointment at a

3. M. Gell-Mann, "A Schematic of Baryons and Mesons," *Physics Letters*, 8, 214–215 (1964).

faculty meeting by passionately arguing that the ace model was the work of a "charlatan."[4]

Gell-Mann and Zweig both agreed on the basic attributes of the quark model. There were three species of quarks. The "up" quark (u) and the "down" quark (d) have spin ½ and isospin ½ (called an *isospin doublet*). A third quark, called the *strange quark* (s), has isospin 0 (called an *isospin singlet*) and "strangeness" one. This third quark is needed to explain the existence of baryons carrying a nonzero strangeness number and also mesons like the positively charged K meson, which is composed of an up quark and an antistrange quark; and the neutral K meson, which is composed of a down quark and an antistrange quark. The three quarks—up, down, and strange—each have what is called *baryon number* equal to ⅓, so that the baryon number of the three quarks constituting a proton or neutron adds up to one.

The particle physics community was skeptical about the idea of protons and neutrons being made up of three quarks, and the need for fractional electric charge did not go down well either. Gell-Mann himself was not convinced that his quarks were real. He treated them as a productive way of visualizing how matter is constituted, and as a mathematical trick to make sense of his Eightfold Way model of particles.

During a visit Gell-Mann made to the University of Toronto in 1974, he told me that he had initially submitted his quark paper to *Physical Review Letters* but it was rejected. He told me that he then phoned the theory director at CERN at the time, Leon van Hove, and said that he was planning to submit his quark paper to *Physics Letters B*, where Van Hove was an editor. Van Hove asked him what a quark was, and he said it was a particle making up the triplet SU(3), which had fractional electric charge and baryon number ⅓. Van Hove was dismissive and didn't think it was a good idea for Gell-Mann to submit the paper for publication. However, Gell-Mann did, and it was accepted and published by *Physics Letters B*.

Yet soon, in 1968, the Stanford Linear Accelerator (SLAC) was used to begin experiments to try to detect quarks by bombarding atomic nuclei with electrons. The idea was to use the latest accelerator technology to repeat the Rutherford experiments that had discovered protons inside atomic nuclei. Rutherford had scattered a beam of alpha particles (helium nuclei composed of two protons and two neutrons) off gold leaf, and had found a sufficient number of alpha

4. George Zweig, "Origins of the Quark Model," in *Baryon '80: Proceedings of the 4th International Conference on Baryon Resonances*, ed. Nathan Isgur, July 14–16, 1980, Toronto. Quoted in Andrew Pickering, *Constructing Quarks: A Sociological History of Particle Physics* (Chicago, IL: University of Chicago Press, 1984), 89–90.

particles scattering at large angles to conclude that they had hit hard objects—namely, the nuclei of the gold atoms. Similarly, in the SLAC experiment, if the angles of the scattering of electrons hitting the insides of nuclear targets were sufficiently large, this would signal a significant deflection of the electrons by hard, massive objects, and would provide at least indirect proof that quarks existed inside protons and neutrons. The experiments did indeed produce a distribution of scattering angles that strongly suggested that there were small objects inside protons.

Two kinds of experiments had already been performed in the linear collider to contribute to this conclusion. One was called *elastic proton–electron scattering*. These experiments showed that the proton was not a pointlike object like the electron; it had a diffused structure. This led physicist Robert Hofstadter to experimental investigations of the structure of the proton during the early 1950s, for which he won the Nobel Prize for Physics in 1961. Subsequently, a series of experiments of so-called *deep inelastic scattering*, in which scattering electrons and protons produced other particles, did not behave as expected with increasing energy. The data suggested that there were some hard objects inside the proton. These experiments were performed on the 22-GeV, two-mile long linear electron accelerator at SLAC beginning in 1967. This research heralded the new "hard-scattering" era of particle physics, compared with the soft-scattering era that had previously dominated experimental particle physics. *Hard* scattering refers to the hard objects inside protons that were hit by the electron beams whereas *soft* scattering refers to proton–proton collisions in which a plot of the scattering cross-sections displayed a rapid fall-off with increasing energy, indicating that these experiments were not investigating the interior of the protons.

Initially, the Stanford experimentalists were not able to explain these radical deep inelastic scattering results. They knew that something important had been observed, but they were unable to understand the implications fully. At that time, particle physicists had not accepted the reality of fractionally charged quarks inside protons and neutrons. James Bjorken, fondly known as "BJ" by his colleagues, who was a member of the theory group at SLAC, analyzed the experimental data and discovered a scaling relationship that could constitute proof of the existence of quarks inside the proton.[5] He analyzed the SLAC data in a somewhat esoteric way, using what was called "current algebra sum rules," and found that the electromagnetic structure of the proton scaled with a scaling parameter that consisted of the ratio of the energy loss by electrons radiating

5. J.D. Bjorken, "Asymptotic Sum Rules at Infinite Momentum," *Physical Review*, 179, 1547–1553 (1969).

off photons and the energy of the new particles produced in the deep inelastic collision.

The experimentalists appreciated that Bjorken had made an important discovery, but a fuller understanding of the experimental results came about only when Richard Feynman visited SLAC in 1968 and invented what he called the "parton model" of strong interactions.[6] The partons were hard objects being hit by the electrons inside the proton, and Feynman was able to explain Bjorken's scaling result using this model. I discuss the parton model in more detail later in this chapter. Eventually, Feynman's partons were identified with Gell-Mann's quarks. Further experiments at SLAC, Brookhaven National Laboratory, and CERN were able to reveal the need for the quarks to have fractional electric charge.

In light of all this experimental evidence, there was a growing impetus for particle physicists to accept that the quarks were real particles and not some fictitious mathematical notion. Gell-Mann was also eventually convinced that his quarks were real. However, why were the fractionally charged quarks not seen as independent particles experimentally? Why were they only able to be detected indirectly inside the protons and neutrons, when electrons bounced off them? In fact, how *were* they confined inside the protons and neutrons? Answering these questions led to an important turning point in the development of particle physics, and eventually the quark model and the parton model were subsumed in quantum field theory based on the gauge principle.

The strange physical contradiction between the quarks and gluons of quantum chromodynamics (QCD)[7] and the old particles of nuclear physics, such as protons, neutrons, and electrons, is that the latter particles can be detected directly either by ionization of nuclei producing electrons or by high-energy accelerators. The fact that one cannot detect quarks and gluons directly as free particles outside the proton was initially considered a failure of the QCD theory. It has not yet been possible to derive a theory of quark and gluon confinement from first principles in QCD. That is, it has not been possible to provide a convincing explanation about why the quarks and gluons appear to be trapped within the nucleus and invisible to detection. Instead of obtaining an explanation of confinement from the basic mathematical formulas of QCD, physicists have been forced to come up with a somewhat ad hoc phenomenological explanation added in by hand. However, in the current particle physics community, this problem is simply ignored, and the hidden quarks and gluons are detected

6. R.P. Feynman, *Photon Hadron Interactions* (Reading, MA: W.A. Benjamin, 1972).

7. QCD is the currently accepted theory of strong interactions involving quarks and gluons, which will be explained in more detail later.

indirectly through hadronizing jets appearing as events after the collisions of protons with protons or with antiprotons. These jets consist of streams of hadrons (the particles that contain quarks and gluons), which, when analyzed, tell you which jet is associated with which quark in the detection process. Although the mystery of confinement has been, to some extent, set aside by the theoretical and experimental particle physics community, it still remains a problem that has to be resolved eventually to make QCD a convincing theory.

TRYING TO MAKE SENSE OF QUARKS

Along with the mystery of "quark confinement," there was another serious problem in understanding the nature of quarks. The proton and neutron were made of different combinations of three up quarks and down quarks. However, these up and down quarks looked identical as far as the quantum spin ½ is concerned. According to Pauli's exclusion principle, which he discovered during the development of quantum mechanics during the 1920s, three fermions of spin ½ (such as quarks) cannot occupy the same quantum state to make up the proton. Something was wrong. It's like attempting to put three unruly identical triplets into a playpen, but each one refuses to be in the same space with the other two. If you try to put three of these up and down quarks into the same quantum state inside the proton and neutron, this would violate quantum statistics and Pauli's exclusion principle. Theoretical physicist Walter Greenberg proposed a way of solving this conundrum. He changed the nature of the quantum statistics for the quarks, calling the new statistics "parastatistics," which allowed one to get around the problem with the Pauli exclusion principle for the quarks. To follow through with the analogy of the triplets, Greenberg proposed to put them in a newly constructed playpen in which they were comfortable with one another. By changing the quantum statistics, he managed to get the three quarks in the same quantum state. However, this was a somewhat radical solution to the problem.

Then, Yoichiro Nambu and Moo Young Han proposed yet another radical idea. They invented a new kind of charge associated specifically with quarks in order to bypass the problem of the Pauli exclusion principle. Murray Gell-Mann and Harald Fritzsch proposed that this new charge be called "color charge." Now, each up and down quark could have one of three "colors"—red, blue, or green. These colors are, of course, not real colors, but are simply used as labels for the new charge. This proposal provided, in effect, three times more quarks for nature to choose from when building particles. The observed hadrons such as baryons and mesons were termed *colorless* or *white*, indicating that their color charges always combined to make the observed hadrons color-neutral.

In collaboration with Swiss physicist Heinrich Leutwyler, Gell-Mann and Fritzsch discovered the mathematical theory underlying the colored quarks. They bound the three colored quarks in the proton by massless force carriers that they called *gluons*, which in turn had their own colors. The mathematical symmetry group describing these colored quarks and gluons was SU(3), but this was a different kind of SU(3) from Gell-Mann's original Eightfold Way SU(3). The different kinds of quarks and leptons are named by their "flavors." The three flavors of quarks known during the late 1960s—up, down, and strange—constituted the fundamental triplet of SU(3), whereas the gluon force carriers were represented by the octet of SU(3). In contrast to the colorless photon of electromagnetism, each gluon is a combination of two colors of the three possible colors, to produce eight colored gluons.

In keeping with the fashion among particle physicists of giving fancy titles to particles and theories, Gell-Mann christened the new theory of colored quarks and gluons *quantum chromodynamics*, using the Greek word *chromo*, which means "color." To accommodate the fact that quarks were always confined and never seen experimentally in high-energy collisions, the hadrons, which are the particles that undergo strong interactions such as protons and neutrons, were colorless and were described as *color singlets* in QCD. In other words, the colored quarks were always combined to produce a colorless hadron. For example, in the mesons, which are composed of a quark and an antiquark, a red quark combines with an anti-red quark to produce a colorless meson. And in the proton, a red, green, and blue quark combine to produce a colorless proton. This means that the hadrons we observe in accelerators do not reveal the colored quarks, which are confined inside them. This, of course, does not explain the dynamics of confinement, and how they came to be that way, but simply indicates the fact that we cannot observe free quarks. The dynamic mechanism of confinement is still a controversial issue in particle physics. (See Figure 1.5 for a summary of the elementary particles of the standard model, and Figure 1.6 for a description of the quark colors.)

FOUR, FIVE, AND SIX QUARKS FOR MUSTER MARK!

James Bjorken and Sheldon Glashow published a paper in 1964 predicting the existence of a fourth quark, which they called the "charm" quark.[8] The four quarks were represented by the fundamental quartet representation of SU(4), which replaced the triplet representation of SU(3). I also came up with the

8. B.J. Bjorken and S.L. Glashow, "Elementary Particles and SU(4)." *Physics Letters*, 11 (3), 255–257 (1964)

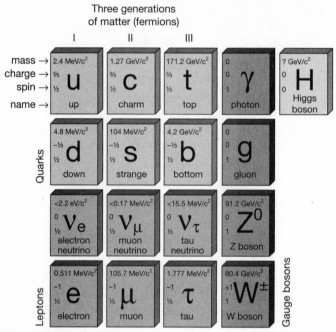

Figure 1.5 The elementary particles of the standard model of particle physics. The quarks and leptons are named by their "flavor," with six quark flavors and six lepton flavors. The scalar Higgs boson is not a gauge boson, so it is not included in the fourth column of gauge bosons.

SOURCE: PBS NOVA/Fermilab/Office of Science/US Dept of Energy.

prediction that a fourth quark should exist, and published a paper in *Physical Review* in 1965,[9] with a fractionally charged fourth quark and fractional baryon number. Bjorken and Glashow's fourth quark had integer charge and baryon number because it was still early days for physicists to accept the idea that quarks were fractionally charged. I called my additional fractionally charged quark simply the "fourth quark," not nearly as flashy as "charm."

The three quarks—up, down, and strange—constituted three so-called flavors of quarks. The high-energy experiments showed two kinds of currents associated with the weak force to create radioactive decay. One is an electrically charged current in which quarks are coupled to the charged intermediate vector boson W, and the other is a neutral current that is coupled to the neutral vector boson Z. In weak decays of particles, the charged W boson associated with the charged current can change the charge and flavor of quarks, while the neutral Z boson cannot change the quark charge and flavor. During these

9. J.W. Moffat, "Higher Symmetries and the Neutron-Proton Magnetic-Moment Ratio," *Physical Review*, 140 (6B), B1681–B1685 (1965).

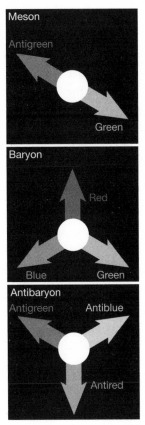

Figure 1.6 The quark color combinations of mesons and baryons.
SOURCE: Wikipedia.

neutral Z interactions, no decays of hadrons have been observed that change the quark flavor through the decay.

This restriction of neutral current decays in the flavors of quarks is known as the *anomaly problem*. In 1970, Sheldon Glashow, John Iliopoulos, and Luciano Maiani used the idea of a fourth quark to solve this anomaly problem in the quark model. They found a way to explain the experimental fact that, in neutral current decays, there was no change of flavor going from the decaying quark to its decay products containing quarks. The *GIM mechanism* (with GIM standing for Glashow, Iliopoulos, and Maiani) explained the lack of flavor-changing neutral currents in weak interactions. This observational fact called for the existence of a fourth quark, which had been predicted in 1964 and 1965.

The quark revolution began to culminate in 1974, when one group at Stanford, headed by Burton Richter, and another at Brookhaven, headed by Samuel Ting, discovered a narrow resonance that was identified immediately as being composed of a charm and an anticharm quark. Ting called the new charm resonance "the J particle," whereas Richter called it "the psi particle," and it eventually became known as the *J/psi particle*. The narrowness of the new resonance was caused by it being a tightly bound composite of a charm and an anticharm quark, which was later called a *quarkonium system*.

We now had four quarks confirmed experimentally: up, down, strange, and charm. In 1977, Leon Lederman and his collaborators at Fermilab discovered a fifth quark, which was called the *bottom* (or *beauty*) *quark*. It was known theoretically from the properties of QCD that there had to be a sixth quark and, of course, it would be called the *top* (or *truth*) *quark*. Indeed, it was discovered in 1995, also at Fermilab. The names *top* and *bottom* had been introduced by Israeli physicist Haim Harari in 1975, two years before the discovery of the bottom quark, to replicate at higher masses the up and down quarks.

The story of these last two quarks, or the third generation of quarks, is worth telling in more detail. In 1973, Japanese physicists Makoto Kobayashi and Toshihide Maskawa had predicted the existence of a third generation or family of quarks to explain the violation of charge conjugation and parity in the decay of K mesons. Charge conjugation is a mathematical transformation of a positively charged particle to a negatively charged one, and vice versa. This transformation turns a particle into its antiparticle. Parity is left–right symmetry in particle physics. This third generation of quarks was required to implement the GIM mechanism. The 1978 discovery by Martin Perl at SLAC of the tau lepton strengthened the need for the introduction of a fifth and sixth quark to implement the GIM mechanism fully, which explained why flavor-changing neutral currents were not detected.

It was not easy to discover these fifth and sixth quarks. Early searches for the top quark at SLAC and at the German accelerator, the Deutsches Elektronen Synchrotron (DESY) in Hamburg, came up with nothing. In 1983, when the super proton synchrotron (SPS) at CERN detected the W and Z bosons, experimentalists at CERN felt that the discovery of the top quark was imminent. A race soon ensued between the Tevatron in the United States, with an energy of 2 TeV, and the SPS collider at CERN to discover the top quark. However the SPS machine reached its energy limit without finding the top quark, which was expected, at the time, to have a mass of about 40 GeV, whereas the energy limit of the SPS was 77 GeV.

At this stage, only the Tevatron accelerator at Fermilab had enough energy to detect the top quark, with a mass that was now expected to be greater than 77 GeV. The two detectors, the Collider Detector at Fermilab (CDF) and the D0 detector also at Fermilab, were actually built to discover the top quark. Indeed, in 1992, the two groups associated with these two detectors saw the first hint of a top quark. It was a tempting clue that the top quark discovery was imminent.

By 1995, there were enough events to establish the existence of the top quark at an energy of about 175 GeV.

We now have six quarks discovered in the lineup of elementary particles. In the standard model, however, there also had to be six leptons to match them. Indeed, new leptons were discovered eventually. In addition to the electron and the muon, there was the tau and also a neutrino associated with each of these leptons: the electron neutrino, the muon neutrino, and the tau neutrino. These quarks and leptons are the basic building blocks of the standard model (Figure 1.5).

THE FORCES OR INTERACTIONS IN NATURE

There are four basic forces in nature that have been confirmed experimentally: strong, electromagnetic, weak, and gravity. The weakest force is gravity. Isaac Newton published in his *Principia* in 1687 the first universal gravity theory based on his mechanics. Newton hypothesized that the inverse square law of gravity held true everywhere in the universe. The strength of this gravitational force was proportional to Newton's gravitational constant, and his calculations agreed with the observed 28-day period of the moon with remarkable accuracy. In 1916, Einstein published his seminal paper on generalizing his special theory of relativity to include gravity and named it the "general theory of relativity."[10] He postulated that gravity was not a "force" between two masses, as it had been under Newton, but was a warping of spacetime geometry by matter. A key prediction of his new theory was accounting for an unusual feature of the planet Mercury's orbit. The perihelion, or closest approach of the planet to the sun, was "precessing," or moving in a rosette shape. Astronomical observations over a century showed that Mercury's orbit was precessing by a tiny anomalous amount that did not fit Newton's theory. Einstein calculated correctly the anomalous precession of Mercury to be approximately 43 seconds of arc per century, which did fit the astronomical data. A second key prediction, of the bending of light by the eclipsing sun, was verified in 1919 by astronomers, including Arthur Eddington. Today we interpret the gravitational force between massive objects as being caused not only by the warping of spacetime in the presence of matter, but also, in the quantized version of Einstein's gravity theory, by the force carrier called the *graviton*. In contrast to the photon, the carrier of the electromagnetic force, the graviton has never been detected experimentally. In

10. Albert Einstein, "Grundlage der allgemeinen Relativitätstheorie," *Annalen der Physik* (ser.4), 49, 769–822 (1916).

quantum gravity, Einstein's geometrical theory of gravity is interpreted as a particle physics force, with the graviton as the carrier of the force.

The second weakest force is the one responsible for the radioactive decay of matter, discovered by Henri Becquerel in 1896, following the discovery of X-rays by Wilhelm Röntgen. Becquerel and his followers, including Pierre and Marie Curie, were interested in radioactive elements such as uranium, which emitted gamma rays (photons) to become other elements or isotopes, such as plutonium. In the language of the standard model of particle physics, we interpret radioactive decay as the weak decay of hadrons, such as neutrons, into leptons, such as electrons, muons, and neutrinos. For example, the neutron can decay into a proton plus an electron and an antineutrino. The carriers of the weak force have been identified as the massive W and Z bosons, discovered at CERN in 1983. Because the W and Z have masses, the weak force is a short-range force, unlike gravity or electromagnetism, which are long-range forces.

The next strongest force is electromagnetism, which is the force that we are most aware of in everyday life. It was originally thought to be two separate forces—electricity and magnetism—but, in 1873, James Clerk Maxwell unified them within his famous electromagnetic field equations. Electricity manifests itself in an inverse square law such as gravity between two electrically charged particles, either positively charged or negatively charged. Magnetism displays itself as two magnetic poles that always come in pairs, a north and south pair. Danish physicist Hans Christian Ørsted in 1820 had discovered that electricity moving in a wire induced a magnetic field that deflected the north and south poles of a compass. Following this, Michael Faraday, during the 1830s, discovered magnetic induction—the fact that electric fields produce magnetic fields and vice versa. This discovery paved the way for Maxwell's electromagnetic field equations. The modern interpretation of electromagnetism finds expression in the theory of quantum electrodynamics (QED), in which the electromagnetic force between charged particles is carried by the photon.

The fourth force in nature is the strong force, which is responsible for binding protons and neutrons in nuclei. It is about 200 times stronger than the electromagnetic force, which in turn is about 10^{38} times stronger than gravity. In the standard model of particle physics, the strong force is carried by the massless gluons, which bind quarks together in hadrons such as protons and neutrons, and thereby bind protons and neutrons together in atomic nuclei.

During the past four decades, theoretical physicists have devoted much effort to attempting to unify the four forces of nature in a grand design. A successful unification would mean that there is ultimately only one force in nature, carried by a massive boson often named the *X particle*, which would reveal itself at very high energies. The electroweak theory represents a partial unification of weak interactions and electromagnetism. The modern attempts at unification have not been as successful as Maxwell's unification of electricity and magnetism. In particular, there has not yet been a successful unification of gravity with the other three forces. During the 1970s and 1980s, there were three main attempts to unify all the forces: grand unified theories (GUTs), supergravity, and string theory.

RELATIVISTIC QUANTUM FIELD THEORY

Early quantum mechanics was not consistent with special relativity theory. Pioneering physicists such as Paul Dirac, Werner Heisenberg, Wolfgang Pauli, Pascual Jordan, and Victor Weisskopf during the 1930s developed a relativistic version of quantum mechanics that was eventually called *relativistic quantum field theory*. Their initial theory quantized Maxwell's classical field equations and was called *quantum electrodynamics*. That is, it made Maxwell's classical electromagnetic field equations consistent with the concept that energy comes in quantum packages the size of Planck's constant multiplied by the frequency of the electromagnetic waves.

QED soon ran into difficulties because the calculation of the energy of the electron interacting with itself, called the *self-energy*, was infinite, which made calculations in QED meaningless. It was not until the advent of renormalization theory during the late 1940s and early 1950s—first proposed by Hendrik Kramers and later developed by Hans Bethe, Richard Feynman, Julian Schwinger, Sin-Itiro Tomonaga, and Freeman Dyson—that a method for removing the infinities in quantum field theory calculations was found.

Back in 1934, Enrico Fermi had developed a theory of the weak force, or beta decay. His theory explained the radioactive decays of nuclei by a simple formula involving effectively just the product of the fields of the leptons (e.g., the electron), including the one neutrino that was expected to exist at the time, and the proton and neutron fields. The constant determining the strength of the weak force was eventually named the *Fermi coupling constant* (G_F), and it described the strength of the interactions of the particles and fields. This theory explained accurately at low energies such transmutations of matter as a neutron decaying into a proton plus an electron and an antineutrino (Figure 1.7). (The lifetime of the free neutron is about 11 minutes.)

As studies of Fermi's beta decay theory continued, physicists recognized that it could not be fitted into the paradigm of a renormalizable quantum field theory, one with calculations of decay reactions of particles that do not produce meaningless infinities. A technical reason for this is that Fermi's coupling

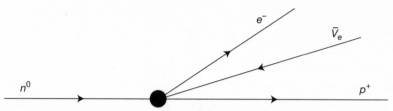

Figure 1.7 Feynman diagram for four-fermion interaction beta decay (a neutron decaying into a proton, electron, and antineutrino)
SOURCE: quantumdiaries.org

constant G_F is not a dimensionless number, but in certain units has dimensions of inverse mass squared. What this means in practice is that calculations of radioactive decay become infinite, and indeed these infinities cannot be removed by the technique of renormalization, as was done successfully for QED, in which the coupling constant, called the *fine-structure constant*, is a dimensionless number. QED's fine-structure constant, alpha (α), which is equal to one divided by 137 and is dimensionless, does lead to a renormalizable quantum field theory, which allows for the calculation of finite quantities such as a scattering cross-section.

Theorists also discovered that if you quantize the gravitational field—interpreting the gravitational force between particles as being carried by the massless graviton—then the theory of quantum gravity is not renormalizable, just as Fermi's theory of weak interactions is not renormalizable. Newton's coupling constant G_N, which measures the strength of the gravitational force, is not a dimensionless number either, but, as for Fermi's constant G_F, it also has dimensions of inverse mass squared, thus rendering any quantum gravity calculations as meaningless infinite quantities.

One of the founders of quantum field theory, Paul Dirac, was never happy with renormalization theory. He did not feel that the cancelation of infinities or ignoring infinite quantities that are supposed to be mathematically small in perturbation theory calculations could be the final answer for quantum field theory. In theoretical physics, we are faced primarily with trying to solve equations that do not have exact solutions. Therefore, we must solve problems by expanding the equations in a mathematical series with terms that are multiplied by successive powers of a small, dimensionless constant. The idea is that when you add up all the infinite terms in the series, you will obtain the exact solution of the equations. The perturbation series only works for dimensionless constants that are less than unity (one). Otherwise, the series will not converge to the exact solution. In particular, if we truncate the series at a certain term, we anticipate that the sum of the terms up to this point gives a good solution to the equations, which can then be compared with experiments.

Dirac expressed his disapproval of renormalization theory with its reliance on perturbation theory:

> Most physicists are very satisfied with the situation. They say: "Quantum electrodynamics is a good theory and we do not have to worry about it anymore." I must say that I am very dissatisfied with the situation, because this so-called "good theory" does involve neglecting infinities which appear in its equations, neglecting them in an arbitrary way. This is just not sensible mathematics. Sensible mathematics involves neglecting a quantity when it is small—not neglecting it just because it is infinitely great and you do not want it![11]

Richard Feynman, famous for his invention of Feynman diagrams for particle physics interactions and for his contributions to the development of QED, said in 1985:

> The shell game that we play…is technically called "renormalization." But no matter how clever the word, it is still what I would call a dippy process! Having to resort to such hocus-pocus has prevented us from proving that the theory of quantum electrodynamics is mathematically self-consistent. It's surprising that the theory still hasn't been proved self-consistent one way or the other by now; I suspect that renormalization is not mathematically legitimate.[12]

Attempts have been made to put renormalization theory on a more rigorous mathematical basis. The published work by Nobel laureate Kenneth Wilson made a significant contribution to our understanding of renormalization theory.[13] His work in condensed matter physics and relativistic quantum field theory further developed what was called *renormalization group theory*, which had been published originally in a seminal paper by Murray Gell-Mann and Francis Low in *Physical Review* in 1954.[14] The work by Wilson made

---

11. Paul A. M. Dirac, "The Evolution of the Physicist's Picture of Nature," *Scientific American* 53 (May 1963). Cited in Helge Kragh, *Dirac: A Scientific Biography* (Cambridge, UK: Cambridge University Press, 1990), 184.

12. Richard P. Feynman, *QED: The Strange Theory of Light and Matter* (Princeton, NJ: Princeton University Press, 1985).

13. K.G. Wilson, "The Renormalization Group: Critical Phenomena and the Kondo Problem," *Reviews of Modern Physics*, 47 (4), 773 (1975)

14. M. Gell-Mann and F.E. Low, "Quantum Electrodynamics at Small Distances," *Physical Review* 95 (5), 1300–1312 (1954).

renormalization group theory more attractive to mathematicians and physicists, although the basic problem of infinite constants such as charge and mass was still present.

ENTER THE W PARTICLE

The electromagnetic force between two electrons or between a positron and an electron is produced by the charged particles exchanging massless photons. In 1938, Swedish physicist Oskar Klein proposed the equivalent of the photon for the weak force. This boson carrier particle came to be known as the *intermediate vector boson*. It is a spin-1 particle that was eventually called the *W boson*. In weak interactions, the W interacts with quarks and leptons, and the strength of this interaction is measured by a dimensionless coupling constant, g. Fermi's coupling constant G_F is proportional to the square of g divided by the square of the W mass. Making the W mass heavy enough makes the weak-interaction theory with the intermediate vector boson W reduce at low energies to Fermi's original weak-interaction theory with Fermi's coupling constant G_F (Figure 1.8).

Because the weak interaction is much weaker than the electromagnetic interaction, the W particle must be heavy, physicists reasoned—between 50 GeV and 130 GeV. Because of the heavy mass of the W, in theory it would decay into other particles such as leptons and have a short lifetime. This means that the range of interaction of the W particle carrying the weak force is very short compared with the infinite range of interaction of the electromagnetic force carried by the massless photon.

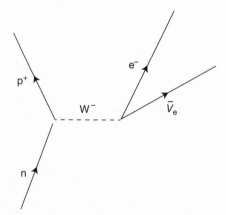

Figure 1.8 Feynman diagram for neutron decay into proton, electron, and antineutrino, mediated by the W boson.
SOURCE: pfnicholls.com

However, it was soon found that the difficulty of obtaining finite calculations in weak-interaction theory was not much improved. The W boson turned out to have similar problems with mathematical infinities as those in Fermi's theory. Quantum field theory calculations for the intermediate vector boson model also produced infinities. In addition, a serious failure of probabilities to add up to 100 percent, in perturbation theory calculations for the scattering of particles involving the W, occurred at an energy of about 1.7 TeV. This meant that the calculated probability of scattering of particles would exceed 100 percent, which of course was impossible.

Such was the situation with the weak-interaction force during the late 1950s and early 1960s. How could one make sense of calculations with weak interactions in quantum field theory? Many attempts were made to resolve this conundrum; but for one reason or another, all failed.

GAUGE THEORY AND THE W PARTICLE

Back in 1918, Hermann Weyl attempted to unify gravity and electromagnetism by generalizing the geometry of Einstein's gravity theory. This geometry is *Riemannian*, invented by Bernhard Riemann during the 19th century, and describes a non-Euclidean curved space, such as the surface of a balloon or Einstein's spacetime. In electromagnetism, the electromagnetic fields are described mathematically as vector fields. These vector fields can be compared in their direction and magnitude at different points arbitrarily distant from each other in spacetime.

Weyl introduced the idea of the "gauge" of a vector field in the geometry of spacetime, similar to the different "gauges" of railroad tracks. In Einstein's spacetime, a locomotive can travel on any railroad track, whereas in Weyl's more generalized spacetime this is not true. In his theory, Weyl was able to unify gravity with electromagnetism by generalizing the Riemannian geometry on which Einstein's theory of gravity was based. In addition to having a metric tensor field, which in Riemannian geometry is used to determine the infinitesimal distance between two points in spacetime, Weyl included a vector field into the geometry. Together, these fields gave rise to both gravity and electromagnetism.

When a vector, a field with a direction in space, is transferred from one point to another distant point, Weyl claimed that it no longer maintained its integrity or its gauge, unlike the case in Einstein's gravity. However, Einstein criticized Weyl's ideas, claiming that nature did not work this way, and eventually Weyl abandoned his theory. The theory simply was not in accord with what you get when you measure the distance between two points in spacetime.

But Weyl did not give up on his idea of gauge in spacetime. In 1929, he attacked the problem again from a different angle. Maxwell's equations of the electromagnetic field remained the same—that is, they are invariant under a mathematical transformation of what is known as the *vector potential field*. This invariance was called *gauge invariance* and had been discovered during the development of QED during the 1930s, soon after Dirac's discovery of his wave equation for the electron. What Weyl discovered is that there is a fundamental connection between the electromagnetic field and the phase of Dirac's wave function for the electron.[15] More is said about gauge theory in Chapter 3.

As we recall, Oskar Klein predicted the existence of the W bosons. He published a paper[16] in which he extended the higher dimensional spacetime unified theory of gravity and electromagnetism of Theodor Kaluza. Later, in 1938, Klein introduced into this theory a massive intermediate vector boson. This was the first appearance in particle physics of what we now call a *nonabelian gauge field*, representing a vector particle. The properties of this nonabelian gauge theory are also discussed in Chapter 3. It must be appreciated that, at that time, during the 1930s, experimentalists had never observed a massive boson with spin 1, which the theorists were predicting.

GENERALIZING MAXWELL'S EQUATIONS

In 1954, two physicists, Chen-Ning Yang and Robert Mills, published their important generalization of Maxwell's equations of electromagnetism.[17] They introduced the idea of isotopic spin space.[18] Instead of the electric charge on particles, they concentrated on the isotopic spin charge of protons and neutrons. When this isotopic spin charge was conserved, then a proton and a neutron would appear to be the same spin-½ particle. When the proton and neutron are subject to strong interactions, they interact as if they are the same particle even though the proton is positively charged and the neutron has zero charge. The small difference in their masses resulting from the interaction of

---

15. The phase of a wave is a particular point in the cycle of a waveform, usually measured as an angle.

16. Oskar Klein, "Quantentheorie und Fünfdimensionale Relativitätstheorie," *Zeitschrift fur Physik A*, 37 (12), 895–906 (1926).

17. C.N. Yang and R. Mills, "Conservation of Isotopic Spin and Isotopic Gauge Invariance," *Physical Review*, 96, 191–195 (1954).

18. "Space" here does not refer to our everyday three-dimensional space, but is a mathematical concept to describe isotopic spin.

the proton with electromagnetism is not important when they interact through strong interactions. This leads to the isospin symmetry of strong interactions.

The idea of isotopic spin space was originally introduced by Werner Heisenberg in 1934. The fundamental paper that Yang and Mills published explored the possibility of the gauge invariance of the interactions of protons and neutrons under isotopic spin rotations in the isotopic spin space. Using the analogy of the gauge invariance of QED and Maxwell's equations, they stated that all physical processes are invariant under local, spacetime-dependent, isotopic spin gauge transformations. Their paper, underappreciated at the time, would become a significant mathematical building block for the future of particle physics.

Ronald Shaw, a research student at Trinity College Cambridge, who was supervised by Abdus Salam, a fellow at St. John's College Cambridge, discovered this mathematical theory of isotopic spin interacting with nucleons (protons and neutrons) independently; it was part of his PhD thesis. Shaw was two years ahead of me in his PhD research when I arrived at Trinity College as a student in 1954. During discussions with Shaw, I learned Salam's opinion about his PhD project. Because, at the time, the physical significance of the isospin generalization of Maxwell's equations was not appreciated, Salam told Shaw that he could not see any future significance for physics in this mathematical development. Unfortunately, Shaw took Salam's advice and did not attempt to publish this idea from his thesis as a paper. If he had published a paper, he would have been as well-known for this significant contribution to physics as Yang and Mills.

Around 1955, Yang gave a talk at the Institute for Advanced Study in Princeton about the theory he had developed with Mills. Wolfgang Pauli was in the audience, and with his usual perspicacity, he asked a question about the presumed mass of the force-carrying W particle because, already at this time, issues had begun to arise about the nature of the hypothesized intermediate vector particle responsible for nuclear radioactive decay. Moreover, Pauli had also speculated on a generalization of Maxwell's equations similar to the Yang–Mills generalization. However, he had not published his results because of his concern about the issue of the mass of the gauge boson that carries the force between isotopic spin charges. *Gauge boson* is a term for a boson associated with a field that satisfies a gauge symmetry (see Chapter 3). This isotopic spin gauge boson eventually became identified with the charged W intermediate vector boson.

Pauli asked Yang early during his presentation, "What about the mass of your gauge boson?" Yang was unable to provide a satisfactory answer. Later during the talk, Pauli again asked the question, "What about the mass of the gauge boson responsible for carrying the force between the nucleons?" Again, Yang was not able to answer the question, but he became so disturbed by this line of questioning that he refused to continue with his talk, and sat down. However, Robert Oppenheimer, who was chairing the seminar, urged Yang to continue his talk, which he did.[19]

As it turned out, Pauli foresaw all the problems that were going to arise in the theory of weak interactions associated with the charged intermediate vector boson because, unlike the photon carrying the electromagnetic force, this boson had a mass. The mass of the charged intermediate vector boson W caused serious difficulties for particle physicists trying to understand the nature of the weak interaction. Almost 60 years after Yang's talk at the institute, how to fit the intermediate vector boson into the standard model of particle physics remains a mystery, particularly if the large hadron collider (LHC) does not confirm the existence of the Higgs particle. The introduction of the Higgs boson and field into weak-interaction theory during the mid 1960s—and later into the unified theory of electromagnetism and weak interactions, electroweak theory—provided a way to resolve the problem of the W boson mass. The so-called *Higgs mechanism* gave the W boson a mass and, in so doing, led to a finite, renormalizable, and self-consistent theory of electroweak interactions.

Despite the experimental successes of QED, some theorists were still not happy with the theoretical foundations of quantum field theory and QED. Such great physicists as Lev Landau, Paul Dirac, and Gunnar Källen did not like having to use renormalization theory to save the calculations of QED from meaningless infinities. Källen and Landau published papers during the 1950s claiming that QED was fundamentally incorrect, because the "renormalization constants" that were used to make the mass and charge in the theory finite were intrinsically infinite, and therefore QED was fundamentally inconsistent as a physical theory.[20]

In their theory, Yang and Mills replaced the photon of electromagnetism with a triplet of bosons that carried the isotopic spin force, one electrically neutral and two oppositely charged. In contrast to Maxwell's electromagnetic theory, in which photons do not interact, or couple, with themselves, the Yang–Mills gauge bosons do interact with each other. If these particles were made massless, the Yang–Mills theory was fully gauge invariant, which means that it could be renormalizable, avoiding the pesky infinities. In their paper, they were concerned that these intermediate vector particles would probably have to have a mass, although they did not explicitly include mass contributions from these bosons in their theory. They recognized that, when you included the protons and neutrons interacting with the triplet of intermediate vector bosons,

19. This scene is described in Robert P. Crease and Charles C. Mann, *The Second Creation* (New Brunswick, NJ: Rutgers University Press, 1996), 194, with quotations from Yang.

20. This was despite the fact that the predictions of Tomonaga, Schwinger, and Feynman in 1949/1950 for QED, such as the anomalous magnetic moment of the electron and the Lamb shift in hydrogen, were confirmed experimentally with amazing accuracy.

then this would introduce a mass dimension into the theory, and the triplet of vector bosons would have to be massive. However Yang and Mills already recognized that putting in the gauge boson masses by hand would break gauge invariance in the theory. This would cause the theory to be beset with infinities, and, unlike QED, it would not be renormalizable.

STRONG INTERACTIONS BEFORE THE QUARK MODEL

During the late 1950s and early 1960s, quantum field theory as the fundamental language of particle physics fell into decline in the high-energy physics community. It was not possible to perform quantum field theory calculations successfully for strong interactions. This was a result of the strong-interaction coupling constant being larger than unity, in contrast to the coupling constant in QED—the fine-structure constant—which was much smaller, equal to approximately one divided by 137. Because of the size of the strong-interaction coupling constant, any perturbation theory calculation using quantum field theory in strong interactions failed, and no known methods in quantum field theory were able to overcome this obstacle at the time.

Theorists turned to the S-matrix approach to strong interactions. Founded by Werner Heisenberg after World War II and also pursued by American physicist John Wheeler, the S-matrix was a way of solving problems in the scattering of strongly interacting particles. The idea was that free, noninteracting particles—hadrons, such as protons—would scatter off other hadrons. The physics of the actual scattering of the particles in accelerators was treated as a "black box"—that is, exactly how the particles scattered was unknowable information to the theorists. This is in contrast to the quantum field theory approach, in which everything should be known about the interactions of the particles at the collision events in spacetime. S-matrix theory became a veritable industry, centered mainly at the University of California at Berkeley, in a group headed by Geoffrey Chew and Stanley Mandelstam. The program led to the phenomenological Regge pole models of strong interactions founded by Italian theorist Tullio Regge.

With the S-matrix approach, it was not necessary to know about the physical processes at the actual scattering site. To calculate scattering cross-sections, it was only necessary to know the incoming particles being scattered and the outgoing particles. The S-matrix was a mathematical way of connecting the incoming particles and the outgoing particles after the scattering had occurred. The idea was that the mathematical properties of this scattering of the S-matrix were sufficient to explain all the needed physics of strong-interaction scattering in accelerators, obviating the need for quantum field theory. Certain mathematical properties called *analyticity* and *unitarity* of the S-matrix played important roles in the development of S-matrix theory. The analyticity had to do with the mathematics of complex variables, and the unitarity had to do with the demand that the scattering processes described by the S-matrix did not exceed 100 percent probability.

The S-matrix program became the dominant activity in the particle physics community during the early 1960s before the appearance of the quark–parton model. An important motivation for this work was the "bootstrapping" mechanism within the S-matrix formalism. Instead of seeking to explain hadron physics in terms of basic building blocks, the hadrons interacted in a democratic way, each with equal importance. The proponents of the S-matrix theory gave up the idea that some particles were more "elementary" than others. All hadrons—such as protons, neutrons, and hadron resonances—these proponents claimed, were created equal, and bootstrapped themselves to produce a self-consistent solution to strong-interaction physics. In other words, exchanges between the particles produced the strong forces that held them together. Experimentalists observed that the spins or angular momentum of the hadron resonances when plotted on a graph lay on straight lines that increased from zero with increasing energy, and theorists conjectured that this resulted in an infinite number of particle spins as one went to higher and higher energies. Those promoting the bootstrapping program claimed that one should not base particle physics on a fundamental particle unit such as the quark, but that the mutual interactions of the hadrons (considered to be elementary particles at the time) bootstrapped themselves to produce other hadrons. All of this was expressed in terms of the S-matrix.

Tullio Regge introduced the idea of Regge theory in 1957. By having the angular momentum or spin of the hadron resonances not take on integer and half-integer values, but take on continuous complex values as functions of energy, Regge identified the hadron resonances as poles in the complex angular momentum plane. Regge poles became a flourishing phenomenological activity during the 1960s and 1970s, occupying particle physicists more than quantum field theory. The quark–parton model eventually became more popular in the particle physics community because it was deemed that the S-matrix approach to strong interactions, including Regge pole theory, was not as promising an approach to understanding particle physics as had been hoped by its pioneers.

GETTING A GRIP ON THE STRONG FORCE WITH THE QUARK–PARTON MODEL

Along with the program of S-matrix theory and Regge poles, physicists pursued other ideas in attempting to understand how the hadrons that had been discovered up until then interacted with one another and how they could be categorized into groups of particles. Physicists such as Abdus Salam and John Ward, Sheldon Glashow, and Jun John Sakurai extended the ideas of Yang and Mills into strong-interaction physics. They used the symmetry group SU(3) to describe the Yang–Mills interactions, which led them to eight intermediary vector particles carrying the strong force, thus extending the Yang–Mills proposal of a triplet of intermediate vector bosons as occurs in SU(2). These eight bosons are the "octet" in SU(3). Yuval Ne'eman, who was a member of Salam's group at Imperial College London, independently published a paper"[21] in which he assigned the known baryons and mesons to octets of SU(3).

A month after Ne'eman's paper, Murray Gell-Mann sent out a preprint of his paper called "The Eightfold Way: A Theory of Strong Interaction Symmetry," which contained independently the same ideas as those proposed by Ne'eman. The octets of SU(3) baryons that both Gell-Mann and Ne'eman were discussing accommodated the "flavors" of baryons and mesons. Flavor, we recall, is a characteristic that categorizes different hadrons. Gell-Mann never published his paper because he doubted whether experiments could verify his proposals. However, he eventually published a paper in which, in Gell-Mann's clever way, he cautiously discussed the possible symmetries of strong interactions between particles.[22] Only in one of the last sections of the paper did he actually discuss his idea of the Eightfold Way in SU(3) as a possible description of strong interactions.

In 1974, when Gell-Mann visited the University of Toronto as my guest, he advised me that when writing physics papers one should attempt to include all possible variants of the proposed theory in the paper, thereby preventing some future author from generalizing your theory and leaving you behind without credit. His paper on the symmetries of baryons and mesons, which was revised twice before it was finally published, was a good example of that philosophy because, in it, Gell-Mann managed to cover the most reasonable ways of grouping hadrons. One of the interesting aspects of the paper is that Gell-Mann appeared to renounce the whole idea of a gauge theory of quantum field theory to describe strongly interacting particles, sticking to his use of group theory to classify the hadrons. The possibility of a gauge quantum field theory only came later, with his proposal of QCD.

Richard Feynman, one of the inventors of QED, which enjoyed remarkable experimental success, wanted to get back into the theoretical game of particle

21. Y. Ne'eman, "Derivation of Strong Interactions from a Gauge Invariance," *Nuclear Physics*, 26, 222–229 (1961).

22. M. Gell-Mann, "Symmetries of Baryons and Mesons," *Physical Review*, 125, 1067–1084 (1962).

physics during the early 1970s. He had been preoccupied mainly with weak interactions and, in collaboration with Gell-Mann, had discovered the V minus A (V – A) theory of weak interactions, which incorporated the violation of parity (left–right symmetry). Robert Marshak and George Sudarshan had discovered independently V – A weak-interaction theory. V – A stands for *vector minus axial vector*, referring to the two currents of weak interactions.

Feynman sought out James Bjorken at Stanford and learned about his theoretical analysis of the detection of hard-core particles inside the proton and neutron at SLAC. Feynman reinterpreted the physical meaning of Bjorken's scaling laws, which described the SLAC results on deep inelastic scattering of electrons and protons, and invented a mathematical way of describing the probabilities of how many of these core particles were embedded in the proton and neutron. Although he knew about Gell-Mann's three quarks being the basis of his SU(3) group theory of the quark model, Feynman wanted to approach the problem from a more general point of view, based on Bjorken's analysis of the experimental data. Despite the fact that Feynman had produced important research in weak-interaction theory in collaboration with Murray Gell-Mann, Feynman and Gell-Mann were quite competitive, and Feynman wanted to demonstrate his own brilliance in the field of strong interactions. He called his particles "partons," and established the probabilities for how many would exist inside the proton for a given energy and momentum transfer in the scattering of protons and electrons in the accelerator.

Feynman conceived of the proton as being filled with many of his tiny, hard *partons*. He assumed that these particles were not interacting with one another—that is, they were freely moving particles inside the proton. He gave no explanation regarding why these free partons did not escape the proton to become visible in accelerators. He pictured the proton as a flat pancake, with the flatness being caused by the contraction of the proton in length because of its moving close to the speed of light (the effect of the Lorentz–FitzGerald contraction of material rods in special relativity).

When one proton is moving past another, you would have two flat pancakes containing partons moving relative to one another at close to the speed of light. He then calculated the probability of a parton in one proton interacting with a parton or partons in another proton when both protons are colliding. This interaction could be electromagnetic as a result of the exchange of photons between the electrically charged partons. Moreover, a parton in one proton could also interact with partons in the other proton through strong interactions by exchanging mesons. From his calculations of the electromagnetic interactions of the partons in the protons, Feynman was able to obtain Bjorken's scaling law and explain the high-energy defractive behavior of the electron/proton cross-sections measured at SLAC. His explanation was much simpler and more intuitive than Bjorken's esoteric explanation based on particle physics and quantum field theory calculations. However, what did this parton model of Feynman's have to do with the quark model, in which the protons and neutrons were made up of three quarks, not *many* quarks? Could the parton be identified with the fractionally charged quarks?

Experiments were then begun at SLAC and CERN to discover how many quarks or partons there were in a proton. The initial results did not prove promising. The theorists were not able to establish that there were only three hard, pointlike objects in the proton. How could they account for the many partons in Feynman's model of the proton? Theorists were not long in coming up with ways of explaining this situation. In an analogy with atomic physics models of atoms, they postulated that the three quarks of Gell-Mann or the aces of Zweig were valence quarks. Analogous to the cloud of electrons surrounding the protons and neutrons in atoms, they hypothesized that there was a "sea" of quarks and antiquarks buzzing around the three basic valence quarks. The experimentalists still did not find results that could identify partons with quarks. Something was missing. So they postulated further that the interactions that bound the valence quarks inside the proton at lower energies created the difficulty in distinguishing how many partons or quarks were inside the proton. Later, when color charge for quarks was invented, and the interaction binding quarks was understood to be caused by colored gluons, the experimental results of proton collisions at CERN gave a comprehensible picture of partons being identified with quarks.

Feynman's parton model eventually became an important technical method for analyzing proton–proton and proton–antiproton collisions. The quark–antiquark interactions that sprayed out as jets from the proton–proton and proton–antiproton collisions were analyzed using his parton distribution functions. Also, Bjorken's scaling laws were reconfirmed experimentally; these techniques were used for both electron–proton scattering and in weak interactions for collisions involving neutrinos.

Two schools of thought emerged during the 1970s concerning progress in high-energy physics. One was that Feynman's parton model was successful in explaining the short-distance scaling behavior of the SLAC experiments because it, essentially, ignored any attempt to incorporate quantum field theory. The other school of thought was that one should incorporate the Feynman and Gell-Mann parton–quark model and the scaling laws into quantum field theory, in particular into a Yang–Mills field theory.

During these investigations, David Gross and his graduate students David Politzer and Frank Wilczek discovered what they called "asymptotic freedom." They found that the partons and quarks became essentially free, noninteracting particles at high energies, because the coupling strength between them decreased as the energy increased. This was in contrast to pure QED of photons and electrons. In the case of QED, as an electron penetrates the cloud of annihilating positrons and electrons surrounding the electron target, the electromagnetic force increases. This is called the *antiscreening* or *antishielding effect*. In contrast, Gross, Wilczek, and Politzer, and also, independently, Gerard 'tHooft discovered that the strong force is screened and decreases in strength as it plows its way through the virtual cloud of partons (or quarks–antiquarks) and closely approaches the target quark. The reason for the difference between pure QED and the quark model is the flavor and color properties of quarks. This asymptotic freedom of quarks was confirmed experimentally; at very high energies, quarks can behave as if they are "free" particles. This discovery strengthened the need to describe strong interactions within the Yang–Mills gauge theory. Gross, Politzer, and Wilczek won the Nobel Prize in 2004 for discovering asymptotic freedom in QCD.

Scattering experiments during the early 1970s confirmed the color charge of quarks. The color charge avoided the violation of Pauli's exclusion principle, which says that you cannot have identical fermions in the same physical state. The numerical factor of three that occurred in cross-section calculations, indicating that there were three "colors"—red, blue, and green—associated with each quark, was found to be real, although, of course, there are no *actual* colors in quarks. There had been a healthy amount of skepticism about the reality of the hypothesized characteristic of quark colors. According to QCD, the rate of decay of a pi meson depends on the color charges of the quark and antiquark that compose it. The rate of decay of an electrically neutral pi meson into two photons is nine times faster with colored quarks than it would be if the quarks were colorless. These predictions of QCD theory were verified experimentally first at the Adone electron–positron collider in Italy. This was a strong impetus for the particle physics community to accept, finally, the fact that although the quarks could never be observed outside the proton and neutron, they truly did exist as real objects, confined inside the proton and neutron.

Originally, Gell-Mann contended that, in studying quarks, one should only be concerned with the white, colorless baryons and mesons, not with what they were possibly composed of. This was because he was still skeptical about the reality of quarks. Physicists have now abandoned this position. Today, when experimentalists at the high-energy accelerators study the debris of particles smashing together, they have to consider the color aspects of quarks and gluons. It is not enough to consider just a colorless gluon. The disturbing psychological aspect of all these colored attributes of quarks and gluons is that you cannot "see" them experimentally. However, the indirect information that emerges through the scattering of protons at energies such as those that are now achievable at the LHC is overwhelming. Seeing is believing, or believing is seeing, as David Gross has said. When two protons collide at the LHC, jets of hadrons are seen to emerge from the point of collision. Experimentalists are now able to interpret these jets as being the quarks and the gluons. The experimental evidence that we have to consider eight colored gluons and not just one colorless gluon, is now substantial and incontrovertible. Software has been developed for the huge grid of computers used by the LHC to analyze the collision data that probe the quark and gluon content of these jets. The jets issuing from the spray of particles after the collisions can be identified as gluons with all their eight color combinations taken into account. The idea of quarks and partons can no longer be considered just a mathematical construction. It has become part and parcel of the everyday life of experimentalists at the LHC.

The remarkable story of the discovery of quarks and gluons illustrates the necessary and fruitful interplay between theory and experiment in physics and in science in general. The imagination and sometimes wild speculation of the theorists are restrained by the straitjacket of experimental physics through the extraordinary efforts of the experimentalists at high-energy accelerators such as SLAC, the Tevatron, and the LHC.

2

Detecting Subatomic Particles

Our understanding of matter has developed through reducing it into smaller and smaller units. Starting with the Greeks, this process has continued through the centuries, except for a lengthy hiatus from the Middle Ages to the 19th century. Einstein's observation of Brownian motion, which indicated the existence of atoms, and then the detection of the electron by J.J. Thomson sped up the whole process of reductionism. The invention of the accelerator meant that particles such as protons were able to collide at higher and higher energies, enabling us to "see" subatomic particles at smaller and smaller distance scales. We will soon be able to probe the structure of matter down to distances of only about 10^{-18} cm, corresponding to an energy of 14 TeV. This should allow us to observe the physical properties of what we now believe are the ultimate units of matter—namely, the quarks and the leptons.

BEFORE MODERN ACCELERATORS AND COLLIDERS

The idea of a particle in the modern sense began with Max Planck in 1900 and Albert Einstein five years later. Planck discovered that the light produced from the walls of a black-body cavity[1] comes in the form of parcels of energy, the amount of which is equal to a constant h times the frequency of the light v. This was a complete surprise in experimental physics at the time. It was widely accepted, and had been proved by James Clerk Maxwell back in the mid 19th century, that light existed as waves. In fact, Planck never accepted the idea that his radical discovery of packets of energy was true, and described nature accurately, nor did many of his colleagues. Einstein, the iconoclast, took Planck's

---

1. A black body is an idealized physical system that absorbs all of the electromagnetic radiation that hits it and emits radiation at all frequencies with 100 percent efficiency.

discovery a step further in 1905 by claiming not only that the black-body walls produced light in quantum packages, but that light itself consisted of quantum packages in the form of particles, which we now call *photons*.

This idea met with opposition even though Einstein figured out a way to reconcile the idea with known experiments, such as one conducted by Philipp Lenard in 1900. In this experiment, light was shone on a metal surface consisting of a cathode, a negatively charged electrode. It knocked out electrons from the metal; the ejected electrons were picked up by another electrode called an *anode*, which is positively charged. This produced a tiny but measureable current. These measurements led to an important observation: the energy and, in turn, the number of electrons knocked out of the metal were not proportional to the amplitude of the light wave hitting the metal, which is what one would expect from waves, but rather were proportional to the frequency of the light wave. Lenard and other physicists at the time could not understand the meaning of these results, because they fully believed that light consisted of waves. Moreover, the effect was color dependent. Light has colors like the rainbow, and only light of blue or violet seemed to have enough energy to eject the electrons. The red, yellow, and green parts of the spectrum were not able to do so.

Einstein figured out that there was an experimentally observed photon energy threshold of about 1.5 eV, at which the light was able to kick out the electrons. Indeed, the blue light had an energy of approximately 3.5 eV, and thus blue and violet light played a special role in this experiment. Einstein understood that the amount of energy left over from 3.5 less the 1.5 threshold energy was carried by the electrons knocked off the metal. From Einstein's theoretical calculations, assuming that the light consisted of particles of energy equal to Planck's $E = h\nu$, the electron energy should increase linearly with the increasing frequency of the light shining on the metal, an important prediction of his particle idea of light. Einstein published these ideas in a 1905 paper titled "On a Heuristic Viewpoint Concerning the Production and Transformation of Light."[2]

It was not until 1914 and 1915 that Robert Millikan, after diligent effort on a series of experiments, proved that Einstein's prediction was correct. Despite this experimental confirmation, Millikan, as well as Planck and Niels Bohr in Copenhagen, still did not believe in Einstein's proposal that light consisted of particles. Since Maxwell, everyone knew that light was made up of waves. After all, light diffracts and goes around corners; light waves interfere with one another constructively and destructively. All of this was well-known

2. Albert Einstein, "Über einen die Erzeugung und Verwandlung des Lichtes betreffenden heuristischen Gesichtspunkt," *Annalen der Physik*, 17 (6), 132–148 (1905).

experimentally. Did we actually have to go back to Isaac Newton's idea that light consisted of "corpuscles"? Did we have to go back to the 17th century to understand the photoelectric effect?

Later, during the 1920s, much progress toward understanding the particle nature of matter was made experimentally with the invention of the *photomultiplier* and the advance of vacuum tube technology. Eventually, after the photomultiplier instrument was perfected, it was able to detect single photons for the first time. With the advance of this "single-photon counting," experiments were able to prove Einstein's principle of the photoelectric effect—that light did indeed consist of particles.

Despite the Millikan experiment and the photomultiplier results, physics luminaries such as Niels Bohr still refused to believe in the idea that light consisted of these photon particles. Indeed, Bohr steadfastly opposed the idea, suggesting that the photoelectric effect was the result of the violation of the conservation of energy. It was not until 1922 that Arthur Compton scattered photons off electrons and proved conclusively that the experimental result of the scattering could only be explained if light was made up of these photon particles. Einstein was awarded the 1921 Nobel Prize in Physics (given to him in 1922) for his ingenious explanation of the photoelectric effect. Eventually, Bohr grudgingly accepted the fact that light did indeed consist of photon particles.

Russian physicist Pavel Cherenkov made an important discovery in 1934, which came to be called *Cherenkov radiation*. What he found was that when he passed charged particles through certain solvents, he saw a blue flash of light when they interacted with the molecules of the material. This was no surprise, for such an effect, called *fluorescence*, had already been observed by, among others, Marie and Pierre Curie in their experiments with radium. However, Cherenkov went further, and discovered that when the charged electrons passed through water in a tank, the electrons were actually moving faster than light in that medium, because light travels slower in water than in a vacuum. As a result of the excessive speed of the electrons, they had to shed energy. This energy was in the form of photons being radiated, which produced a bluish cone the size of which depended on how many photons were produced as the electrons raced through the water. The theoretical understanding of Cherenkov's experiments was achieved by Il'ia Frank and Igor Tamm. These two theorists, together with Cherenkov, won the Nobel Prize in Physics in 1958 for discovering and explaining Cherenkov radiation.

Photons can be detected in other ways. When a photon passes a nucleus with a sufficiently strong electric force (Coulomb force), it can produce spontaneously an electron and positron (positively charged electron) pair, which can be detected. If an electric field is strong enough, it can reverse the process and make an electron and a positron annihilate and produce a photon. This phenomenon is called *spontaneous creation* of either photons or pairs of electrons and positrons.

Another effect is that when an electron passes through the electric field of a nucleus, it produces a spray of photons from the interaction of the electron with the electric field, which can also be detected. This effect is called *Bremsstrahlung*, which in German means the slowing down or "braking" of the electrons by the electric field. The Bremsstrahlung effect can repeat over and over again with different nuclei and, eventually, with sufficiently high energy, a shower of photons is produced. These showers are characteristic effects caused by high-energy photons and electric fields.

The electrons and positrons produced by the photons near an electric field make "tracks" that can be observed in other experimental devices. The earliest instrument invented to follow and measure such movements of ionized particles was the "cloud chamber," invented by Scottish physicist Charles Wilson in 1895. Wilson discovered that condensation occurs around a charged particle in a medium of water vapor. Indeed, the way to measure the charge of the ions is by counting the vapor droplets that surround the ions in the cloud chamber. Through diligent experimental research, Wilson eventually perfected his cloud chamber in 1911 by passing the charged particles through water, where the vapor condensing on them produced tracks of bubbles. However, Wilson's cloud chamber was not able to detect the scattering of particles because water is not dense enough to serve as an effective target.

The next major advance in cloud chamber technology came 40 years later. In 1952, Donald Glaser invented the "bubble chamber" at the University of Michigan. He started with a medium of diethyl ether, the solvent and anesthetic, only 3 cm^3 in volume, but soon produced chambers of much larger sizes. He sent beams of charged particles through the medium, which produced tracks in the form of bubbles. Glaser also used liquid hydrogen kept at a very low temperature of –253°C. The hydrogen atoms, of course, contained a proton nucleus, so the proton became a target for the charged particles. Now experimentalists were able to produce significant scattering events using charged particles and protons. The bubble chamber played an important role in the development of accelerators, culminating in the giant European bubble chamber at CERN, called *Gargamelle* (Figures 2.1 and 2.2).

Bubble chambers produced clear tracks of charged particles, which made it possible for experimentalists to study nuclear scattering processes more efficiently (Figure 2.3). Bubble chambers served as physicists' favorite particle detectors for many years. They were also filled with liquid freon, liquid propane, and other liquids that provide sufficiently dense targets for the scattering of particles. For his important contribution to experimental particle physics, Glaser was awarded the Nobel Prize for Physics in 1960.

Figure 2.1 Old Fermilab bubble chamber.
SOURCE: Wikipedia Commons, from Fermilab.

Figure 2.2 CERN's Gargamelle bubble chamber. © CERN

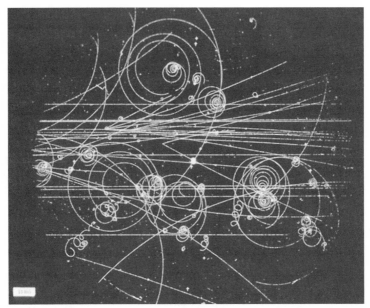

Figure 2.3 CERN photograph of early bubble chamber tracks. They are produced in a small bubble chamber filled with liquid hydrogen. The photo shows a pion beam striking a proton target in the upper left, which creates a spray of new particles. © CERN

MODERN DETECTORS

Significant technological advances have been made in the instruments used to detect subatomic particles at the colliders and accelerators that smash particles together at high energies. One of the earliest such detectors was the *spark chamber*, invented by Shuji Fukui and Sigenore Miyamoto. It was important in particle physics research from the 1930s to the 1960s. The spark chamber consisted of a stack of metal plates immersed in a gas such as helium or neon. When a charged cosmic ray particle passed through the detector, it ionized the gas, and sparks were generated between the plates, showing the path of the cosmic ray particle.

Another type of detector was the *proportional wire chamber*, invented in 1968 by CERN experimentalist Georges Charpak, for which he won the Nobel Prize in 1992. The idea was based on the existing *wire chamber* detector that had been used at accelerators to detect particles of ionizing radiation, which itself had been a significant advance over the Geiger counter[3] and *proportional counter*. Like the spark chamber, these detectors all use metal, high voltage, and a gas medium to track the path of ionized particles.

---

3. A Geiger counter is a portable ion counter used to measure radioactivity.

In high-energy accelerators, tracking the path of a particle is necessary to measure its energy or mass, charge, and other properties. This was done for a long time, up to the early 1960s, using bubble chambers and taking photographs. However, with the rapid development of electronics, it became possible to devise a system with a fast electronic read-out of results. By tracking the path of a particle in a magnetic field, one can measure the charge on the particle and its mass, and then infer other quantum numbers possessed by the particle to identify it.

In the case of the proportional wire chamber, and most other detectors of charged particles, the particles are led through a uniform magnetic field that turns their paths into spirals because of the electromagnetic force on the fast-moving charged particles. By measuring the curvature of the spirals, we can determine the charge and momentum of the particle. The accuracy of the reconstruction of the particle's path can be greatly increased by observing that the ions need time to drift toward the nearest wire. The measurement of this drift time can determine where the particle passes the wire. This apparatus is called a *drift chamber*. By combining many drift chambers with wires orthogonal to one another, and by arranging the orthogonal wires to be orthogonal to the beam direction of the particles, a very accurate measurement can be made of the position of the particle, and from this we can determine its momentum. From the momentum of the particle we can tell what kind of particle we are dealing with.

One of the great advances that came with the proportional wire chamber is that it was now possible to be selective in identifying the particle one was searching for from the myriad of possibilities after a collision. This cuts down the number of events possible in a given measurement to a number of events that can be handled more easily by the computer systems used to analyze the data. Indeed, many millions of potential "events" occur when protons collide with protons or antiprotons, and a selective triggering mechanism is required to reduce the number of possible events to a manageable size. Richard Feynman once famously said that colliding protons together was like bashing two full garbage cans; one needed a method of sorting out the debris.

Another detection instrument, which is currently used at CERN, is the *calorimeter*. It is a composite detector that absorbs particles to measure their energy. The trillions of interactions generate showers of particles, which is why the name *shower counter* is also used for a calorimeter. Most of the particle energy in the showers will be converted into heat for this kind of detector, which explains the name calorimeter (*calor* means "heat" in Latin), but in practice, no temperature is measured in these detectors. Particles enter the calorimeter and initiate a particle shower. The particles' energy is deposited in the calorimeter, where it is collected and measured. A total measure of the energy can be obtained by a complete containment of the particle shower, or the energy can be sampled. There are two kinds of calorimeter: the electromagnetic calorimeter and the hadronic calorimeter. The former is used to measure the energy of particles that interact electromagnetically whereas the latter is designed to measure particle energies that interact via the strong nuclear force. An important feature of the electromagnetic calorimeter is that it is the only practical way to detect and measure electrically neutral particles, such as photons, in a high-energy collision.

A calorimeter is made of multiple individual cells, over the volume of which the absorbed energy is measured. The cells are aligned to form towers typically along the direction of the incoming particles. Incoming electromagnetic particles, such as electrons and photons, are fully absorbed in the electromagnetic calorimeter, which takes advantage of the comparatively short and concentrated electromagnetic shower to measure the energies and positions of these particles (e.g., neutral pi mesons decaying into photons). In a hadronic calorimeter, the incident hadrons (such as protons) may start their showering in the electromagnetic calorimeter, but usually are absorbed fully only in later layers of the apparatus.

The compact muon solenoid (CMS) and a toroidal LHC apparatus (ATLAS) detectors at the LHC have special electromagnetic calorimeter detectors used to measure decays of particles into photons and electrically neutral bosons such as the Z boson (Figure 2.4). Experimentalists are able to detect a light neutral Higgs boson, which can decay into two photons. This particular decay is difficult to measure because the cross-section, or area of interaction, of the two-photon decay is very small. However, the electromagnetic detectors are amazingly sensitive to the amount of light or photons emitted by the Higgs boson in its decay. This experiment is one of the important means of detecting the elusive light Higgs particle, because there is very little background of strongly interacting hadrons such as bottom and antibottom quarks. This calorimeter detector was crucial for the presumed detection of the standard-model Higgs boson, announced on July 4, 2012, by the CMS and ATLAS groups at the LHC.

How do we know what kind of particle we are seeing in a detector? By determining the amount of energy and momentum of a particle, which can be measured in a detector, we can find out the quantum numbers of the particle, such as its charge, mass, and quantum spin, which identify it. Take, as an example, a particle with an energy of 200 GeV and zero momentum. This is a particle at rest. Except for a stable particle such as the proton and electron, a particle can decay into other particles, each individually of smaller energy. This whole process of decay must satisfy the conservation of energy and momentum.

Figure 2.4 Calorimeter at the ATLAS detector. © CERN for the benefit of the ATLAS Collaboration

Normally, the probability of a particle decaying into other particles is determined when the particle is at rest. Some particle decays are very rapid. For example, an electrically neutral Z boson can decay into a positron and an electron with a decay time of about 10^{-24} seconds. This particle has a very short track in a bubble chamber compared with the much longer decay time of a muon, which takes about a millionth of a second (10^{-6} seconds) to decay. The muon, therefore, traverses a sufficient distance in a bubble chamber that its path can be identified as a muon. On the other hand, the distance traversed by the Z boson is so short that the paths in the bubble chamber cannot be used to identify the Z. Instead, we have to identify the specific decay products, the particles produced by the decay of the Z, and determine that the energy and momentum of the decay products equal those of the parent particle. This is possible because of the overall conservation of energy and momentum of the process.

When one is trying to discover a new particle, such as the Higgs boson, one determines the mass and momentum of the particles in the decay products and infers whether the measured results match the physical properties of known particles. If they don't, then it is possible that one has discovered a new particle. If a particle such as the neutral Z boson (Z^0) is being investigated, it will not leave the track of a charged particle. However, one sees a V shape either in the bubble chamber or in the data tracking from modern detectors, signifying that it has decayed into a positron and an electron; this V shape is a signature of the neutral Z boson.

EARLY ACCELERATORS AND COSMIC RAYS

The very first accelerator was used by J.J. Thomson in 1897 when he discovered the electron. This primitive apparatus accelerated particles between two electrodes with different electric potentials. Thomson used electromagnetic fields to determine the ratio of charge to mass of the accelerating particles that were in the form of cathode rays, or negatively charged particles. In contrast to physicists working at modern accelerators, Thomson studied the properties of the beams of particles, not the impact of the beams on a target. Although modern accelerators are huge in size and much more complex than Thomson's accelerator, they work on much the same principles. Accelerators were found commonly in the living rooms of the 1950s and 1960s. Instead of today's flat-screen liquid crystal display TV, the first television sets used a cathode ray tube. The TV tube accelerated electrons, which hit the screen and were deflected, thereby emitting light. The TV with its accelerator was, of course, a tiny version of modern-day accelerators. It produced, at most, a 10,000-eV energy during the acceleration of the electrons.[4] Compare 10,000 eV with the LHC, which has a maximum energy of 14 TeV.

The biggest accelerator is our own universe. The universe produces cosmic rays, the discovery of which in 1910 is credited to Austrian physicist Victor Hess and to Theodor Wulf, a Dutch Jesuit priest and physics teacher. Wulf constructed an electroscope, an instrument that was sensitive to charged particles. He discovered that his electroscope lost energy because it was absorbing radiation. It was originally assumed that this radiation emanated from the earth. But when Wulf asked French physicist Paul Langevin to verify this assumption by doing an experiment with the electroscope on top of the Eiffel Tower, Langevin discovered that the electroscope discharged faster at the higher altitude of the Eiffel Tower than lower down on the ground. The radiation seemed to be coming from the sky.

Then, Hess investigated this phenomenon by ascending in a balloon to see whether the electroscope discharged more significantly as the balloon increased its altitude. Above a height of 5 km, the discharge effect was much stronger than

4. An electron has an energy of 1 V when it is accelerated by an electric field with a potential difference of 1 V.

on the ground, from which Hess concluded that the source of radiation was the universe itself.

American physicist Robert Millikan decided to investigate this phenomenon, too, and named the radiation *cosmic rays*. Perhaps because of this catchy name, Millikan became better known for the discovery of cosmic rays than Hess, Wulf, or Langevin. However, in 1936, Stockholm awarded Hess one-half of the Nobel Prize for discovering cosmic rays, and the other half went to Carl Anderson, who discovered the positron. For reasons unknown, Wulf got nothing.

Recent evidence from spacecraft experiments indicates that the sources of cosmic rays are violent supernovae explosions. The cosmic rays consist primarily of protons. As they come in from the universe, they are accelerated to very high energies, up to 10^{21} eV, which is more than a hundred million times higher than the energies that man-made accelerators such as the LHC can reach. These cosmic rays travel enormous distances through space and finally hit the earth. However, on their way through space, they collide with particles and produce secondary and tertiary cascades of particles. In these cascades, one can detect the positively and negatively charged pi mesons, K-mesons, and muons. It was through using cosmic rays that Anderson discovered the positron in 1932, confirming Dirac's prediction of antimatter.

One of the problems with using cosmic rays as one would use proton beams in accelerators for particle physics experiments, however, is that they bring with them a huge, unknown background. The signal-to-noise ratio is too small to allow a proper control of cosmic ray experiments, compared with experiments performed in the laboratory on earth. Nevertheless, cosmic rays are used today to perform experiments on neutrinos, to determine the oscillation of neutrino flavors into one another, and, potentially, to determine the masses of neutrinos. Present-day ongoing experiments with cosmic rays attempt to identify new particles, despite the serious background problems, and also to obtain an experimental understanding regarding whether Einstein's special relativity is valid at the very high energies of cosmic rays. So far, no experiment has detected conclusively a violation of special relativity theory.

MODERN ACCELERATORS AND COLLIDERS

We now turn our attention to the modern machines that speed up particles very close to the speed of light. When such particles collide with one another, they produce cascades of particles with properties that can be detected by those devices that we have examined earlier. The word *accelerator* signifies generically a machine for accelerating particles to higher energies, whereas the word *collider* denotes the head-on collision between two particles or a particle and an antiparticle. Particles can also be accelerated and then hit a fixed target, consisting of a material containing nuclei. In practice, the terms *accelerator* and *collider* are used interchangeably.

Accelerators come in two basic kinds: linear and circular. The first linear accelerator prototype was built by Norwegian engineer Rolf Wideröe in 1928, and it became the progenitor of all high-energy particle accelerators. He used a method of resonating particles with a radio frequency electric field to add energy to each traversal of the field. Wideröe published the details of his invention in an article in 1928.[5]

Ernest Lawrence studied this article and used the information in it as a basis for building the first circular cyclotron accelerator during the early 1930s at Berkeley. This cyclotron consisted of a round metal chamber separated into two pieces, with a perpendicular magnetic field in the chamber. The two halves were connected by an electric field voltage. Protons were injected into the middle of the chamber and the magnetic field controlled their circular orbits, keeping them in curved paths. Manipulating the electric field allowed the protons to be accelerated as they passed from one half of the instrument to the other. After the protons were accelerated sufficiently, they were removed from the apparatus and focused on a target. This cyclotron was the grandfather of all circular accelerators, including the LHC today. Lawrence received the Nobel Prize in 1939 for developing the cyclotron (Figure 2.5).

The first modern linear accelerator was constructed by John Cockcroft and Ernest Walton during the early 1930s; they received the Nobel Prize in 1951 for splitting the atomic nucleus. Their machine accelerated protons to about 700,000 electron volts (700 keV). This, of course, is low energy compared with present-day accelerators, but was sufficient to study nuclear physics. Only lower energies are needed to probe the protons and neutrons of the atomic nuclei, as opposed to the subatomic quark constituents of the proton and neutron that are studied today. The Cockcroft–Walton accelerator was a small linear accelerator.

The most famous example of a linear collider is the one built at Stanford University during the mid 1960s: the Stanford Linear Accelerator Center (or SLAC). This 2-mi-long collider, the largest linear accelerator in the world, consists essentially of two linear accelerators with the beam in one accelerator going in one direction, and the beam in the other going in the opposite direction. The beams consist of positrons in one beam and electrons in the other

5. R. Wideröe, "Über ein neues Prinzip zur Herstellung hoher Spannungen," *Archiv für Elektrotechnik*, 21 (4), 387–406 (1928).

Figure 2.5 Lawrence's original 11-in cyclotron. Photo courtesy of the Science Museum (London) and the Lawrence Berkeley National Laboratory.

beam. These particles do not experience significant energy loss, referred to technically as *synchrotron radiation*, because the particle trajectories are kept linear and not bent. However, this statement is not strictly true, because at SLAC the beams are bent in a slightly curved shape to force the particles to collide head-on.

Other linear electron–positron accelerators have been built in Italy, Siberia, and France, and a new one is planned for CERN, to be called the international linear collider (ILC). Despite the fame and the usefulness of SLAC and other linear accelerators, currently this type of accelerator is used mainly to give an initial boost of acceleration to particle beams as they enter a main circular accelerator. In the future, however, the very large linear accelerators that are planned, such as the ILC, will be able to obtain more precise data than the circular accelerators because they will be colliding positrons and electrons, reducing the amount of hadronic background that plagues cyclotrons, including the LHC. They will be able to measure electromagnetic particle collisions and to determine precisely the properties of newly discovered particles such as the Higgs boson.

During the years after World War II, the cyclotron, or circular accelerator, was engineered into a machine that accelerated protons to higher and higher energies, culminating in 1983 with the Tevatron machine at Fermilab near Chicago, which accelerated protons and antiprotons to an energy of almost 2 TeV. Cyclotron machines were also built at Brookhaven National Laboratory and at Argonne National Laboratory near Chicago.

The idea of the cyclotron is an important concept underlying all future accelerators because it can accelerate protons from low energies in circular orbits to very high energies, keeping them in tight orbits using magnetic fields. Keeping these tight orbits stable requires what is called *strong focusing*, and this demands building special magnetic field apparatus. The machines designed for strong focusing are called *synchrotrons*, and they have played an important role in the history of accelerators. For example, the Alternating Gradient Synchrotron at Brookhaven and the Proton Synchrotron (PS) at CERN boost the energies of the accelerators at these sites. The round synchrotrons have gaps along the perimeter, where electric fields are applied to accelerate the protons or antiprotons continuously. At some point, strong magnetic fields are used to extract the protons so they can be focused on a nuclear target. Most synchrotrons use protons as the accelerated particle because protons lose less energy in the form of synchrotron radiation when their orbits are bent by magnetic fields. In comparison, electrons lose a lot of energy and therefore are less valuable as initial "kicker" accelerating particles. The energy loss of the particle produces a deceleration, which is what we try to avoid happening.

In 1961, Italian physicist Bruno Touschek developed the first collider with a single storage ring. In this collider, oppositely charged particles are accelerated within the storage ring in opposite directions and then are focused to collide at a point, producing a cascade of particle debris. The energy produced when two particle beams collide in the storage ring is much greater than the energy obtained by a single particle beam hitting a fixed target. The reason is that when a particle hits a fixed target, a lot of kinetic energy continues on after the collision, reducing the amount of energy that can be studied that represents actual events produced by the collision. With the colliding beams, on the other hand, there is much less loss of energy in the collision, making the Touschek storage ring collider a valuable advance in the development of accelerators and colliders.

In the ring colliders, one beam of particles is injected in one direction and the other beam is injected in the opposite direction. The same electric and magnetic fields can be used to accelerate, for example, electrons and positrons, which have opposite electric charges. The applied electric field accelerates electrons and positrons in opposite directions, and during this acceleration they bend in opposite directions in the same magnetic field. In ring colliders, one can use protons in both beams or protons and antiprotons. Recall that these particles can then be accelerated to much higher energies than the electrons and positrons.

The most famous circular electron–positron collider was located at CERN during the 1990s and was called simply the *large electron–positron collider* (LEP). It reached a maximum energy of 209 GeV, which was the combined energy of the two beams at collision, each having an energy of 104.5 GeV. LEP was closed down in 2000 to make way for the world's largest ring collider, the large hadron collider (or LHC). In a ring collider such as LEP, the beams of the particles and antiparticles are kept separate until they are made to cross at certain intersection points. The PS collider at CERN is also such an intersecting ring collider, and in both machines, ingenious technological inventions were required to produce sufficient luminosity or beam intensity at high enough energies.

In quantum mechanics, particles can be described both as particles and as waves, the physical behavior of which can be determined by probability theory. That is, the probability of the particle being at a certain point in space and time can be calculated using quantum mechanics. Short-wavelength particles can be used to examine particle properties at extremely small distances. However, for longer wavelength particles, the experimental results become coarser. The length of the wave of these particles is inversely proportional to their energy, which means that a proton, for example, could be either short wave or longer wave, depending on its energy. A short-wave particle has a higher energy than a longer wave particle.

In quantum mechanics, the wavelength of the particle is inversely related to its energy or momentum. In atomic physics, this energy is in the range of thousands of electron volts (an atom has a scale of 10^{-8} cm). At the scale of an atomic nucleus, the energy is in the millions of electron volts. To study the physics of particles such as quarks inside atomic nuclei, we need energies exceeding billions of electron volts. We use accelerators and colliders to produce particles such as short-lived quarks that do not exist as stable particles on their own in a natural state. However, as we know, we still cannot "see" individual quarks in accelerators, but only indirectly as hadronizing jets in detectors.

American physicist M. Stanley Livingston produced an impressive plot in 1954 showing how the laboratory energy of particle beams produced by accelerators has increased over time (Figure 2.6). This plot has been updated to account for modern accelerators and, remarkably, the energy has increased by factors of 10 every six to eight years. This amazing energy increase is the result

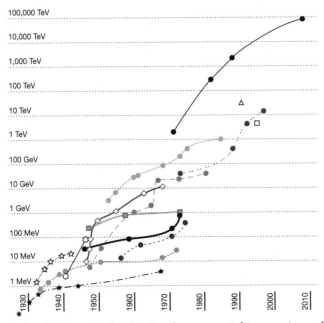

Figure 2.6 Updated Livingston plot showing the exponential increase in accelerator energies over time. When explaining the units, the Snowmass report stated: "Energy of colliders is plotted in terms of the laboratory energy of particles colliding with a proton at rest to reach the same center of mass energy." This is why the collision energy at the LHC appears to be almost 100,000 TeV on the graph. Adapted from 2001 Snowmass Accelerator R&D report; graphic by *symmetry* magazine.

of the ingenuity of physicists and engineers creating new technologies, which in turn allow for the construction of bigger machines.

Yet, increasing the energy of accelerators is not the only significant factor in building these machines. We need higher energies to reach beyond the horizon of known physics, but there are other important factors involved in our study of particle physics. For example, we need beams of colliding particles with a high intensity or luminosity—that is, with a large number of particles accelerated per second. When a particle hits a target, it produces particle reactions that are measured by a cross-section. This cross-section is the effective area of a material composed of nuclei that acts as a target for the beam of particles hitting it, and it is the area containing the resulting particle debris from the collisions that provides the data for the experiment. The important number for physicists is the actual size of the cross-section that contains the reactions. This final number will be the product of the beam intensity, the density of target particles, the cross-section of the actual reaction being produced, and the length of target material that the colliding particle can penetrate.

Another important number is the so-called *duty cycle* of the machine, which is the amount of time that the accelerator can actually be running during the course of a period of experimental investigation. Modern accelerators do not produce a continuous flow of particles because, for one thing, the electric power required would be huge and unsustainable. Thus, the accelerators run as experiments with pulses that go on and off; consequently, the circulating beams pulse on and off as well. If the duty cycle is too short, the cross-sections will not be large enough for data analysis.

A major problem in making sense of cross-section data is the background debris produced by a collision of particles either in a collider or on a fixed target. This background determines the signal-to-noise ratio, in which the noise is the unwanted background of particle collisions that obscures the signal we are trying to isolate, such as identifying a new particle. These backgrounds have to be modeled by large computer simulations and then subtracted from the data to produce a large enough signal-to-background noise ratio to be useful. The background is one of the most important and difficult obstructions to identifying new physics. As the energy and intensity of the particle beams in the accelerators increase and reach those of the LHC at its maximum running energy of 14 TeV, the background noise increases enormously. Very sophisticated computer codes are required to simulate the backgrounds so they can be subtracted during the final analysis of the collisions.

During the 1950s and 1960s, the designing and building of accelerators was a major undertaking around the developed world. Many new machines were added to the roster, producing higher and higher energy collisions of particles, and major new discoveries in particle physics were made. A big fixed-target machine was built at Brookhaven in 1952, with an energy of about 1 GeV. In 1957, the Russians built a machine with an energy of 7 GeV at Dubna, near Moscow. These were followed at CERN by the PS at 23 GeV and then the SPS at an energy of 400 GeV. The DESY, begun in Hamburg, Germany, in 1959/1960, accelerated electrons up to an energy of about 7 GeV, and was instrumental in confirming QED. Important figures in machine building were Wolfgang Panofsky, who helped build SLAC, and Robert Wilson and John Adams. Wilson began building Fermilab in 1967, and Adams built both the PS and the SPS at CERN. Wilson was lab director at Fermilab for several years, and Adams was CERN's director from 1971 to 1980. The SPS at CERN discovered the W and Z bosons in 1983. Simon van der Meer, who had invented a method for accumulating intense antiproton beams, which were made to counterrotate and collide with protons, and Carlo Rubbia, the head of the project, won the Nobel Prize in 1984 for this discovery. The Fermilab machine accelerated protons up to an energy of 500 GeV, and eventually, using superconducting magnets, that machine, called the *Tevatron*, succeeded in doubling this energy to about 1 TeV per beam, or 2 TeV in total energy. The Tevatron had 900 superconducting magnets along the ring, with huge currents running through their magnetic coils, and the currents could reach about 5,000 amp. It was important to keep these magnetic coils at extremely low temperatures to reduce their electrical resistance to zero. This was done by using liquid helium at –270 K. The Tevatron, which among other things, discovered the top quark in 1995, was decommissioned in September 2011.

THE LARGE HADRON COLLIDER AT CERN

We now come to the largest collider ever built, the LHC at CERN. It took 25 years of planning and construction to build this accelerator, which was paid for by a consortium of countries.

The accelerator tunnel, 27 km in circumference, lies between 50 m and 175 m beneath the earth's surface, straddling the French–Swiss border. This concrete-lined tunnel is 3.8 m wide, and was constructed between 1983 and 1988 for the LEP machine. The LHC is an intersecting ring collider, which contains two parallel beam pipes that intersect at four points, and each of the 27-km-long, circular beam pipes contains a proton beam. These two beams travel in opposite directions around the ring. They are kept in circular orbits by means of 1,232 dipole magnets, and an additional 392 quadrupole magnets focus the beams, maximizing the chances of collisions between the protons at the four intersecting points. Most of the magnets weigh more than 27 tons each. It requires 96 tons of liquid helium to keep the copper-clad niobium–titanium magnets at their operating temperature of 1.9 K (–271.25°C). This amazing cooling system comprises the largest cryogenic facility in the world, which is kept at liquid helium temperatures.

When the LHC is running at maximum energy, once or twice a day the protons will be accelerated from 450 GeV to 7 TeV. To accomplish this acceleration, the magnetic fields of the superconducting dipole magnets are strengthened from 0.54 to 8.3 tesla, which is a unit of measure of the strength of magnetic fields. When they are fully accelerated, the protons will be moving extremely close to the speed of light—namely, at about 0.999999991 times *c*, the speed of light. This incredible speed is merely 3 m/s slower than the speed of light. The protons go around the 27-km ring in less than 90 microseconds, completing 11,000 revolutions per second (see Figures 2.7 and 2.8 for diagrams of the LHC and its detectors).

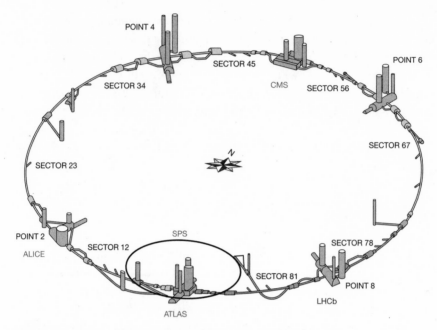

Figure 2.7 Diagram of the LHC showing the positions of the four detectors: ALICE (A large ion collider experiment), ATLAS, LHCb (large hadron collider beauty), and CMS. © CERN

The LHC is actually a chain of accelerators. The protons, which represent the "hadrons" in the machine's name, are prepared before being ejected into the main accelerator by a series of systems that increase their energy. The linear accelerator 2 (LINAC 2), a linear accelerator at the beginning of the LHC, generates 50-MeV protons and feeds them into the proton synchrotron booster, which accelerates them to 1.4 GeV and then injects them into the PS, where they are accelerated to 26 GeV. The SPS is then used to increase the energy of the protons to 450 GeV, and at last these protons are injected into the main ring.

In September 2008, the LHC was finally switched on and shot 2,808 bunches of protons around the ring. Each bunch contained 100 billion protons, and they collided 40 million times per second inside four detectors. The interactions between the two proton beams take place at intervals never shorter than 25 nanoseconds (a nanosecond is 10^{-9} seconds, or 0.000000001 second). The detectors are situated at the intersection points in the accelerator ring: the ATLAS detector, the CMS detector,[6] the LHCb (large hadron

6. ATLAS stands for "A Toroidal LHC apparatus," and CMS means "compact muon solenoid."

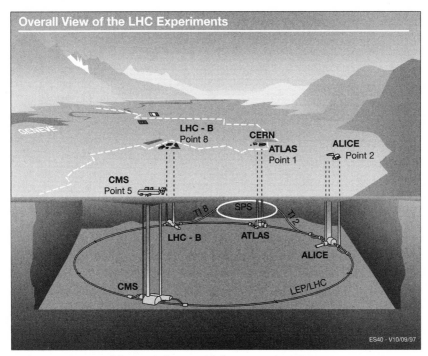

Figure 2.8 Drawing of the LHC above and below ground. © CERN

collider beauty) detector, and the ALICE (a large ion collider experiment) detector. The circulating beam energies of the protons rise to an energy of 7 TeV, and when they collide, they generate a combined energy of 14 TeV, making the LHC the most powerful accelerator ever built[7] (Figure 2.9 shows the ATLAS detector).

However, nine days after the machine was switched on, one of the superconducting magnets blew up because an unstable cable connection evaporated during a high-current test. One of the serious problems in dealing with this accident was interrupting the machine, which was running at a temperature of 1.9 K, using 130 tons of liquid helium. Fixing the magnet was a sensitive operation that took a long time.

As we recall, one of the important elements in an accelerator is the luminosity, or intensity, of the colliding beams. At full luminosity at the LHC, the particle collision rate is about $10^{34}/cm^2/second$. That creates a lot of debris to

7. The superconducting supercollider in Texas, the construction of which was canceled in 1993, was to have had a maximum energy of 40 TeV.

Figure 2.9 The ATLAS detector. Note the person standing at the base, for scale. © CERN for the benefit of the ATLAS Collaboration

sort through! Currently, CERN is upgrading the LHC to produce much higher luminosities and collision rates, which will enhance the possibility of finding new forces and particles. This upgrade of the LHC, to be completed in 2015, will be called the high-luminosity LHC (HL-LHC).

The main experimental program at the LHC is based on proton–proton collisions. However, one month is set aside every year to perform heavy-ion collisions, mainly with lead ions. They are first accelerated by the LINAC 3 linear accelerator, and the low-energy ion ring is also used as an ion storage system. This experiment aims to produce the quark–gluon plasma that is thought to be the initial stage of matter in the early universe. The detector used to study heavy ion collisions is the ALICE detector.

The main purpose of the LHC is to probe the constituents of the standard model of particle physics and to look for exotic phenomena such as mini black holes, extra dimensions of space, supersymmetric particles, and especially the Higgs boson. If the current suggestions of the Higgs boson are confirmed by the HL-LHC, that discovery will reveal the mechanism underlying electroweak symmetry breaking. It will reveal the mechanism that generates the masses of the elementary particles, if indeed a Higgs-type mechanism is responsible for this. If the new boson discovered at the LHC turns out not to be the standard-model Higgs boson, however, then physicists hope that the LHC will

"All right, pal, I'm just saying, that's what I'd do if it was my Large Hadron Collider."

Figure 2.10 © Edward Steed, New Yorker Magazine

be able to identify another mechanism responsible for the symmetry breaking in the electroweak theory.

The LHC is also investigating the nature of dark matter, the dominant and invisible substance postulated to exist throughout the universe to account for the data that show stronger gravity in galaxies, clusters of galaxies, and cosmology than is predicted by Einstein's general relativity theory. The dark matter particles that the LHC is looking for are stable supersymmetric partners of the standard-model particles. So far, no supersymmetric particles have been found at the LHC, including any that could be identified as dark matter.

The building of the LHC was accomplished by an extraordinary consortium of countries contributing to the $9 billion cost of the project. With the machine running for some time, it was hoped that a major new discovery such as the Higgs boson or detection of a supersymmetric particle could be announced. Experimentalists at CERN worried that no new major discovery would be made before the proton–proton collisions ended in late 2012, when the machine would turn to heavy-ion collisions and a two-year period of maintenance and upgrading. In view of the 2011/2012 financial crisis in Europe, such a null outcome could mean that the funding of the machine would be reduced, perhaps leading to another superconducting "supercollider" (SSC)-like setback for particle physics. However, with the discovery of a new boson at 125 GeV, announced on July 4, 2012, with properties consistent with a Higgs boson, a new era of particle physics has begun, and the future running of the LHC seems secured.

3

Group Theory and
Gauge Invariance

To understand particle physics, it is necessary to explain some of the underlying mathematics; otherwise, a deeper insight into the subject is not possible. Many of the basic features of particle physics are mathematically abstract. Because particle physics—as opposed to gravity or other macroscopic physical phenomena—takes place at extremely small distance scales, we use quantum mechanics and quantum field theory to explore the interactions of particles and fields.

SYMMETRY AND MATHEMATICAL GROUPS

A notion that has played a dominant role in classical physics as well as quantum mechanics and particle physics is symmetry. We can describe fairly easily what is known as the symmetry of objects. Take, for example, a solid cube. Every cube has the same features—namely, six square faces and eight corners. All the faces and corners look the same. If you draw an axis through the cube and rotate it 90 degrees around this axis to return where you began, then you get back the same object—an identical face of the cube. This symmetry property of a cube is called an *invariance*, and the rotation is a *transformation* on the cube. The pyramid is another object with faces and corners that are the same as the faces and corners of other pyramids. These objects belong to the class of regular polyhedrons.

However, in classical physics and particle physics, we have to broaden our understanding of symmetry or lack of symmetry, because we are not dealing with the symmetry of physical objects. Instead, we are considering something much more abstract—the symmetry or lack of symmetry of the laws of nature. Humans have a desire to see symmetries in nature. From ancient astronomers up to Johannes Kepler during the 16th and 17th centuries, astronomers insisted on the orbits of the planets being circular, because the circle and sphere were perfect symmetric shapes. However, as Kepler discovered, the orbits of the planets are not circular orbits; they are ellipses. That is, they do not exhibit rotational invariance (i.e., symmetry). The elliptical orbits of the planets depend on a direction in space—the major axis of the ellipse points in a certain direction—and therefore they cannot be rotationally invariant. Isaac Newton discovered from the equations of motion contained in his laws of mechanics and gravity that both circular and elliptical orbits were solutions of his equations. However, if the solutions of Newton's equations of motion were purely circular orbits, then these solutions would describe orbits that are symmetric—independent of direction in space.

Physicists often talk about "broken symmetries," particularly in particle physics. In terms of the previous example, the elliptical orbits of the planets are considered broken symmetries compared with the concept of circular orbits. To understand a broken symmetry, we have to define the initial meaning of a symmetry before we break it. This is done mathematically through what is called *group theory*. The mathematical system of transformations that can reveal the symmetry or invariance of a law of nature is called a *group*. Group theory was developed first by mathematicians, who were not concerned with applying their results to the laws of nature. Only later did physicists adopt the mathematical results to understand symmetry in nature.

The mathematical notion of groups was invented primarily by Evariste Galois, a young French mathematical genius who, during his short lifetime in the 19th century, discovered the properties of groups while investigating solutions of algebraic equations. Norwegian mathematician Niels Henrik Abel, in his similarly short lifetime, at age 19 derived the important result that there is no algebraic solution for the roots of quintic, or fifth-order, equations, or indeed of any polynomial equation of degree greater than four, if one uses just algebraic operations. In theoretical physics, the adjective *Abelian*, derived from Abel's name, has become so commonplace in the physics literature that the initial "A" is now often written in lowercase. Galois proved independently the same result as Abel. To carry out the proof, they both invented the mathematics of group theory.

Another Norwegian mathematician, Sophus Lie, born in 1842, developed the mathematics of group theory further and invented what we call *Lie groups* (pronounced "Lee" groups). A Lie group describes a continuous set of transformations, in contrast to a discrete or discontinuous set of transformations. For example, the rotation of the cube through 90 degrees can be done in a continuous way, from zero to 90 degrees, or in a discontinuous way by breaking up the transformation into intervals of, for example, 20 degrees, then 30 degrees, and finally 40 degrees. Let's investigate these concepts—abelian and nonabelian groups, group theory, and Lie groups.

A mathematical "group" combines pairs of elements in a set using an associated operation or rule. For example, the operation could be addition; you choose a pair of whole integers and add them together to get another whole integer. The resulting number then belongs to the original set, or group, of whole integer numbers. In other words, whole numbers form a mathematical group. Equivalent operations on a pair of elements would include subtraction, multiplication, and division, whether the elements are whole numbers or fractions. However, there are restrictions on the definition of a group.

A group must satisfy four criteria or axioms of group theory. The first is that the elements must be "closed" when the operation on a pair of elements is performed. In other words, the elements close in on themselves; they cannot become anything through the given operation other than members of the group to which they belong. For example, if you add 4 and 6, you get 10. You don't get 10½ or 10.7. Ten is a whole number that belongs to the original set of whole numbers.

The second axiom is the "associative rule." It says that if you have three elements, a, b, and c, you can combine them either as (a + b) + c or as a + (b + c) and they will give the same answer. Similarly, you can combine (a × b) × c or a × (b × c) and you will get the same answer; that is, it doesn't matter what the order of the operations is. One might consider that combining b + c first and then adding a could give a different answer, but it doesn't. In physics, the associative rule has not been considered of great significance, since most of the physical applications of group theory assume that the associative rule is always true.

The third axiom of group theory is that the group has an "identity" element for a given operation. This means, for example, that the identity "I" is such that I × a = a. This rule says that for two elements, a and b, we have a × b = I, where I is the identity element.

This leads to the fourth axiom of group theory, that there exists an "inverse" for every element of the group, using the concept of group identity. That is, b is equal to I/a, where b = I/a defines the inverse of the operation a × b. Thus, this fourth axiom states, in effect, that every element in the set or group must have an inverse.

An important characteristic of group theory is that if you take two elements, a and b, and multiply them together, then a × b may or may not be equal to b × a. This is called the *rule of commutation of elements*. Indeed, matrices in mathematics do not necessarily commute when multiplied together. A matrix is a mathematical abstraction consisting of a block of symbols in rows and columns. When Heisenberg developed his first quantum mechanics, he did not understand that when deriving experimental spectral lines from his new quantum theory, he was using matrices that did not commute. His professor, Max Born, pointed out this important fact when Heisenberg described his new quantum mechanics to him. Here is where Niels Henrik Abel left his indelible mark on mathematics and physics; the groups for which the commutative law holds—that is, for which $a \times b = b \times a$—are called *abelian groups*, whereas those that do not satisfy the commutative law are called *nonabelian groups*.

As we will discover in the following chapters, our understanding of the standard model of particle physics depends critically on the mathematics of group theory—and on one of the two basic kinds of groups in group theory. We will find that a significant part of the standard model is based on nonabelian groups, those in which the elements of the group do not commute. These groups have a continuous set of elements and are a type of Lie group. The elements within them could be representations of particles or particle fields. On the other hand, groups containing a finite or discrete set of elements are commonly used in chemistry and studies of crystals and crystallography.

DEEPER INTO GROUPS

Let us consider some special cases of continuous groups, such as the group of rotations known as O(2). This is the rotation group of a two-dimensional plane, which leaves invariant a circle inscribed on a sheet of paper. Visualize an arrow pointing from the center of the circle to the perimeter of the circle. Like a combination lock, this arrow can be rotated clockwise or counterclockwise through an infinite number of angles about the fixed central point. The order of the operations of rotation is not important; that is, the order of rotations commutes. Therefore O(2) is an abelian group.

Another continuous group is U(1), in one complex dimension. This dimension is described by a complex number $a + ib$, where i is the imaginary number equal to the square root of -1. U(1) is the group of complex numbers for which the product of two complex numbers is also a complex number, and hence they form a group under ordinary complex multiplication. These two continuous groups, O(2) and U(1), are, in fact, the same because the complex number $a + ib$ is actually composed of two real numbers, a and b, such that the one complex dimension is equivalent to the two "real" dimensions, a and b, of O(2). U(1) is designated the *unitary abelian group* in one dimension.

We now step up the number of dimensions to three, so that we have three axes x, y, and z. From these three-dimensional Cartesian coordinates, we can form the continuous group O(3). Picture a rectangular box, or a book, aligned with the x-, y-, and z-axes. We can rotate this object about the x-, y-, or z-axis. It now turns out that a 90-degree rotation around the x-axis, the width of the book, followed by a 90-degree rotation around the y-axis, the height of the book, does not give the same orientation as rotating 90 degrees around the z-axis, the thickness of the book, and then another 90 degrees around the y-axis. It follows that the operations of rotation do not commute. This means that O(3) is a nonabelian group.

An interesting feature of the group O(3) is that we can use a finite number of elements to generate an infinite number of resulting elements, a feature that places this group in the class of Lie groups. Thus, in a Lie group, instead of considering the infinite number of rotations of the box around the x-, y-, and z-axes, we consider just the finite number of "generators" that produce the infinite number of elements of the group. In O(3), there are three generators—namely, the rotations measured by the angles θ_x, θ_y, and θ_z which are related to the three axes x, y, and z. The reason that O(3) is a Lie group is because the rotations are not discrete; they are continuous.

In contrast to O(3), we can also visualize the group O(2) × O(2) × O(2), consisting of three copies of the group O(2). This group also has three generators (angles) that can produce the infinite number of elements of the group. However, in contrast to O(3), the order of rotation of a pair of angles commutes, and therefore this three-dimensional group forms an abelian group, which was not the case for O(3).[1]

We now move to a two-dimensional group in the complex plane, SU(2). Similar to our construction of U(1) in one complex dimension, we can now form the group SU(2) in two complex dimensions. SU(2) stands for *special unitary Lie group in two complex dimensions*. In contrast to the case of O(2) and U(1), SU(2) is not identical to O(3). SU(2) is the group of 2 × 2 complex matrices. If you multiply two of these matrices in one order, you get a third matrix, but if you multiply them in the opposite order, you get a different matrix. As in the case of O(3), SU(2) is a nonabelian group. Similar to O(3), the Lie group SU(2) has a finite number of generators—namely, three—that can generate the infinite number of elements in SU(2). Although SU(2) and O(3) are different Lie groups, they do share the same Lie group algebra, which describes the order or sequence of pairs of rotations. The identity operations for O(3) and SU(2) are different (the identity operation is the one that takes the group back to itself under rotations). For O(3), it only takes 360 degrees of rotation to get back to

---

1. O(3) is the three-dimensional rotation group of all rotations around the origin of three-dimensional Euclidean space. The group SO(3), a subgroup of O(3), does not include reflections as O(3) does, so rotations reflected in a mirror look the same.

the original state, whereas it takes 720 degrees of rotation in O(3) to get back to the corresponding original SU(2) state.[2] Because of this difference, SU(2) is said to have two degrees of rotation such that, in physical terms, SU(2) can describe a particle's quantum orbital angular motion as well as spin, the spin being understood to be the quantum mechanical spin or degree of freedom.

The nonabelian Lie group SU(2) plays a special role in particle physics and in the standard model in particular. It was this group that Yang and Mills used to describe the mirror symmetry of protons and neutrons; under the influence of strong interactions they look exactly alike, and both are designated as *nucleons*. Yang and Mills generalized Maxwell's electromagnetic equations to reach this result. They generalized the U(1) group invariance of Maxwell's equations for electromagnetism to an SU(2) group invariance of the equations. Yang and Mills called the SU(2) description of nucleons the "isotopic spin group."

We can, in fact, increase the number of dimensions of the complex space being acted on by the special unitary groups from 2 to N, thereby forming the special unitary group SU(N), where N equals any number greater than one. In particular, the nonabelian group SU(3) is the group of 3×3 complex matrices. This group forms the basis of the theory of strong interactions called *quantum chromodynamics*. Before the development of QCD, Murray Gell-Mann and Yuval Ne'eman used SU(3) to describe the many new particles discovered during the 1960s. The number of generators of the Lie group SU(3) is eight, corresponding to the eight independent rotation angles in three complex dimensions. (For the group U(3), there would be nine generators.) This is why Gell-Mann called his theory of particles the *Eightfold Way*, alluding whimsically to the Buddhist path to enlightenment.

INVARIANCE: ABSOLUTE OR RELATIVE?

We recall that symmetry in nature means that under a set of transformations, the laws of nature remain the same, or invariant. A fundamental invariance is the one discovered by Galileo. It says that the laws of physics and the results of experiments are the same when expressed in an inertial frame of reference, defined to be one that moves at a uniform speed, when compared with another inertial frame of reference also moving at a different uniform speed. We have experienced this when sitting in a stationary train, looking out the window, and watching another train moving past us slowly. We get the strange feeling that

---

2. In group theory, SU(2) is technically referred to as the *double cover* of O(3).

the train opposite is not moving but we are moving, even though we are not. In Galileo's day, he envisioned ships when thinking about this invariance.

Newton was well aware of this symmetry, called *Galilean invariance*, and used it to construct his laws of motion. He postulated the existence of an absolute space and an absolute time, and incorporated Galilean invariance by saying that objects move relative to the absolute space and that the simultaneity of events in that space was absolute. Nineteenth-century physicists such as Maxwell believed that there had to be a medium in space that carried electromagnetic waves. This medium, called the "ether," in effect would represent Newton's absolute space, because bodies in space were moving relative to the absolute frame associated with the ether. This was an analogy with Newton's notion that bodies move with respect to the frame of his absolute space. At that time, scientists believed that the speed of light was not independent of the motion of the source of the light. That is, light would move at different speeds relative to the absolute reference frame of the ether.

The null experiment of Michelson and Morley, performed in 1879 to detect (finally) the "ether," showed, surprisingly, that the motion of the earth did not affect the speed of light. This indicated that the speed of light was a constant, independent of its source, and therefore there was no absolute frame of reference associated with an ether. The idea of an all-pervasive ether in space was so entrenched in the physics of the 19th century that physicists such as Hendrik Lorentz attempted to explain the Michelson–Morley experiment and its implications for light speed by inventing an electrodynamics of electron interactions. In this theory, Lorentz claimed that the matter making up the interferometry rods of the experiment contracted along the direction of the motion of the earth around the sun, canceling out the putative effect of the ether on the light signal, and thereby saving the idea of the ether.

In 1905, Einstein postulated that the speed of light is constant with respect to the motion of every inertial frame and independent of the motion of the source of light. He discovered a new symmetry of nature in the form of his special relativity theory, which overthrew the older Galilean invariance based on Newton's postulate of an absolute space and time. Now clocks and measuring rods appeared to observers to be affected by their motion. Clocks slowed down as they were accelerated to the speed of light, and measuring rods shrank in size as they were propelled close to the speed of light. (Einstein ignored the idea of the ether.) Moreover, Einstein's postulates in special relativity, including the postulate that the speed of light is constant with respect to all inertial frames, led him to discover that energy could be converted into mass times the square of the speed of light.

German mathematician Herman Minkowski expressed brilliantly the invariance symmetry of special relativity—contained in the group transformation laws discovered by Hendrik Lorentz in 1904—by picturing spacetime as four-dimensional with three space dimensions and one time dimension. The invariance of the laws of nature with respect to the Lorentz transformations is described by the group SO(3,1). This Lie group is a special orthogonal group of transformations in three spatial dimensions and one time dimension. This group replaced the more restrictive symmetry or invariance group associated with what is called the group of Galilean transformations in which the speed of light was treated as an infinite quantity. SO(3,1) describes the transformation of two coordinate reference frames moving with constant relative speed to one another in special relativity. The Galilean group of transformations can be considered approximately true for slowly moving objects such as the horses and carriages or ships of that era.

In classical physics as well as particle physics, the continuous symmetry invariance of equations implies the existence of a conservation law in a theory. German mathematician Amalie (Emmy) Noether derived the basic conservation laws of energy and angular momentum from the translational and rotational symmetry of the field equations of any physics theory. From her work we can derive the conservation of electric charge from the gauge invariance of Maxwell's field equations. Gauge invariance is one of the most fundamental invariance symmetries of nature. Maxwell's equations can be written in terms of a potential vector field, having direction, in four spacetime dimensions. From this vector field, we can derive the strength of the electromagnetic field for Maxwell's equations. A certain transformation in this field, expressed as a derivative of a scalar field with one degree of freedom, leaves the electromagnetic field strength invariant. The derivative, an arbitrary gradient added to the vector potential, does not change Maxwell's field or equations. This transformation of the vector potential and the resulting invariance of Maxwell's equations constitute the gauge invariance of Maxwell's equations. The vector potential field has no absolute values. Although it cannot be detected physically in classical experiments, it does have a deep physical significance in modern gauge theory.[3]

GAUGE SYMMETRY AND GAUGE INVARIANCE

The important symmetry called *gauge invariance* plays a fundamental role in particle physics. When famous mathematical physicist Hermann Weyl published the first unified theory attempting to unify gravity and electromagnetism,

---

3. The vector potential in electromagnetism can be detected according to quantum theory by means of the Bohm-Aharanov phase prescription.

he used the idea of gauge symmetry to try to unify these forces, applying it to spacetime in his unified theory. In addition to the spacetime metric of Einstein's gravity theory of 1916, Weyl introduced a vector field, generalizing the Riemannian geometry used by Einstein to construct his theory of gravity.

In his book, *Space, Time, Matter*, originally published in 1918,[4] Weyl reconsidered Einstein's original purely *metric* theory, in which distances between points were measured using Pythagoras's theorem. Weyl questioned the implicit assumption behind the metric theory of a fixed distance scale or "gauge." What if the direction as well as the length of measuring rods, and also the unit of seconds in measuring the time of clocks were to vary at different places in spacetime? He used the analogy of railway gauges varying from country to country. It was well known to travelers at that time that they sometimes had to change trains at national borders because of incompatible railway track gauges. To create a similar situation in physics, which allowed him to incorporate the electromagnetic vector potential into spacetime geometry, Weyl generalized the coordinate transformation symmetry of Einstein's gravity theory to a gauge symmetry associated with the electromagnetic vector field. By this means, he was able to create his unified theory.

Einstein dismissed Weyl's ideas about gauge symmetry of spacetime. He showed that Weyl's theory would mean that clocks from different parts of space time would tick at different rates when they were brought together to the same place in space. In fact, this would also contradict our understanding of atomic spectral lines, because they would vary from one position in space to another. However, Weyl prevailed much later, in 1929, by publishing an article that applied his ideas of gauge symmetry to the phases of the quantum mechanical wave.[5]

Prior to Weyl's paper, in 1927, another theorist, Fritz London, applied Weyl's early ideas about gauge theory to the quantum mechanical phase of the wave function in Schrödinger's wave equation.[6] In quantum mechanics, we have the duality of wave and particle, as explained by Bohr's complementarity principle, in which he claimed that the wave and the particle are simply two complementary manifestations of matter. An important physical feature of wave motion is its phase. Two oscillating waves described by x(t) and y(t) have a frequency ν, an amplitude A, and a phase θ (Figure 3.1). When the crest and trough differ by

4. Hermann Weyl, *Space, Time, Matter* (Dover Books on Physics), 4th edition (Mineola, NY: Dover Publications, 1952).

5. H. Weyl, "Electron und Gravitation," *I.Z. Physik*, 56, 330 (1929).

6. F. London, "Quantenmechanische Deutung der Theorie von Weyl," *Z. Physik*, 42, 375 (1927).

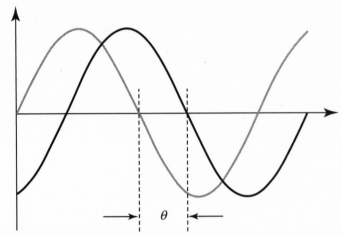

Figure 3.1 Phase shift in waves. An angle (phase) on the horizontal axis increases with time.
SOURCE: Wikipedia Commons

a phase θ of anything other than 180 degrees, then the waves are said to be "in phase" (Figure 3.2). If the phase θ is changed by a specific numerical amount of π or 180 degrees, then the wave motion of y(t) or x(t) is unchanged and the crest of one wave matches the trough of the other wave; they cancel out and the waves are said to be "out of phase" (Figure 3.3).

In quantum mechanics and particle physics, the conservation of electric charge plays a significant role. When waves change phase, they can cause a violation of this conservation of electric charge because the electric charge is the source of electromagnetic fields and waves. However, as shown by London and Weyl, the electromagnetic field arranges itself so that it always cancels out this difference in phase, retaining the conservation of electric charge, which is equivalent to the gauge invariance of Maxwell's equations.

We have now connected a symmetry of quantum mechanical (wave) equations with a conservation law—namely, the conservation of electric charge. We

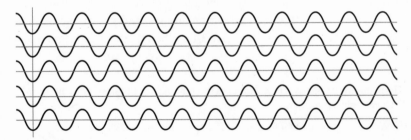

Figure 3.2 Waves in phase.
SOURCE: Wikipedia Commons.

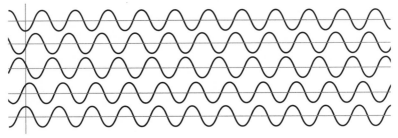

Figure 3.3 Waves out of phase.
SOURCE: Wikipedia Commons.

see that the gauge invariance associated with quantum mechanical wave motion as described by London and Weyl is connected intimately to Emmy Noether's discovery of the connection between symmetries in nature and conservation laws. We can indeed say that electric charge is the source of the electromagnetic field. This interplay between quantum mechanics and particle physics has been important in the development of QED. It can be applied to Dirac's relativistic wave equation for the electron or any other spin-½ charged particle such as quarks.

We must now distinguish between global gauge invariance, or global symmetry, and local gauge invariance. In the case of global symmetry, the changes of phase of the wave equation are the same at every point in spacetime, and this guarantees the global conservation of electric charge. On the other hand, for local gauge symmetry, the phases can change from one point to another in spacetime, but the electromagnetic field always counteracts these changes of phase in such a way that electric charge is conserved. That is, although the waves are out of phase locally from one spacetime point to another, the conservation of charge guarantees that any difference in the phases is canceled, thus preserving gauge invariance.

The concept of local gauge symmetry has played a crucial role in the development of particle physics theories and the standard model. In their nonabelian gauge theory based on SU(2), Yang and Mills extended the idea of phase differences and conservation of charge by replacing electric charge with isotopic spin charge. We recall that the isotopic spin charge is applied to the strong-interaction theory of protons and neutrons. There is a symmetry between the isotopic spin of protons and neutrons in strong-interaction theories that leads to the conservation of isotopic spin charge. Instead of the photon being the carrier of the force between protons and neutrons, as it is in electromagnetism, Yang and Mills invented a vector field B with spin 1 to do that job. In their original article published in *Physical Review* in 1954, Yang and Mills explained that the B-vector meson had to have a charge and to be massless, to mimic the photon in QED. Yet, there are no known electrically charged particles that are mass-less, including the photon, and this fact was troubling for Yang and Mills. It turned out to be a serious flaw in their argument for a nonabelian extension of Maxwell's equations. Despite this, they felt that the idea of this local gauge symmetry and the conservation of isotopic spin through the force field of the B-vector meson was beautiful and elegant, and justified publication in *Physical Review*. As we will see in later chapters, the B-field was replaced subsequently by the color charged gluon particle in the nonabelian SU(3) strong-interaction theory QCD. The gluon is the carrier of the strong force in the atomic nucleus. It is massless, and the electric charge in the theory is replaced by color charge. Moreover, for the weak interactions, the B-field of Yang and Mills's original theory became the charged, massive W boson.

Theorists working on gauge theory up to the present have been focusing on how to interpret the vector field in the nonabelian gauge theory published ini-tially by Yang and Mills for SU(2). In 1983, the model of weak interactions being mediated by two massive vector bosons was confirmed by the discov-ery of the massive W and Z particles. Already during the early 1960s, before the experimental discovery of the W and Z particles, the question had been raised: Where do the masses of the W and Z come from? It seemed unusual that these force carriers should have masses, because the photon and the gluon did not. This search for the origin of the W and Z masses and the masses of the quarks and leptons led to the idea of broken gauge symmetry. Physicists conceived of an initial massless symmetry phase of all the elementary particles, including the W and Z particles, in the early universe, which was then broken spontaneously, thereby producing masses for the elementary particles, except for the photon and the gluon.

The mathematical concepts of group theory, symmetry, and gauge invariance led eventually to the development of the standard model of particle physics based on the group SU(3) × SU(2) × U(1), where the SU(3) sector describes the strong interactions whereas SU(2) × U(1) is the electroweak sector of the model.

4

Looking for Something New at the LHC

The main goal of the LHC is to search for the last remaining undetected elementary particle in the standard model of particle physics, which plays a fundamental role in its framework— namely, the Higgs boson. It was important for the experimental runs in 2012 to build up the luminosity or intensity of the proton collisions so that the new data could produce a more significant statistical signal of the new particle resonance that had been discovered in 2011. That new resonance might or might not turn out to be the Higgs boson.

According to the standard model, the Higgs boson is a massive, electrically neutral particle. It has spin 0, which means that it is a scalar boson field; it does not have any directional properties in space, as is the case with the other bosons. Bosons are the carriers of the force fields between matter particles such as quarks and leptons. The bosons that have been discovered so far are the photon, the gluon, and the W and Z bosons. The Higgs boson has possibly been discovered, whereas others—like the graviton and the inflaton—are still hypothetical.

Historically, since the first accelerator, which was built in 1932, no one has ever detected an elementary particle that has quantum spin 0 and is a scalar field. (In quantum physics, the terms *particle* and *field* are interchangeable.) The pi meson (pion), discovered in 1947 by Cecil Powell, César Lattes, and Giuseppe Occhialini, does have spin 0, but it is not an elementary particle because it is composed of a quark and an antiquark. If the Higgs particle, as an elementary particle, is detected at the LHC, then it would be the first elementary spin-0 particle discovered in nature, with the same quantum numbers as the vacuum.

Of course, if a Higgs boson of spin 0 does exist, it could be a composite of quarks and antiquarks like the pi meson, rather than an elementary particle. However, the pi meson, together with other light mesons, are "pseudoscalar particles"—that is, they have negative parity, in contrast to the elementary Higgs boson, which has positive parity and is a scalar particle. Spin-0 composite scalar mesons have been observed in quark spectroscopy. They are composites of bound quark–antiquark pairs and exist in excited states above the ground state of hydrogen-like systems. The hints of a putative Higgs boson at 125 GeV at the CMS and ATLAS detectors could be interpreted conceivably as a quark–antiquark state. However, this would require the existence of a quark with a mass of about 50 to 60 GeV. No known quark with this mass has yet been detected. If indeed the Higgs boson is an elementary particle with a mass of about 125 GeV, it would be the first elementary particle with spin 0 and positive parity discovered.

Despite its traditional experimental status as a potentially nonexistent elementary particle, the Higgs boson has played a significant role in particle physics and cosmology. It is not as if theoretical physicists have much choice in the matter. Rather, the physics of both the standard model of particle physics and of cosmology demand that the basic Higgs particle be an elementary spinless particle. And it is not the only spin-0 scalar elementary particle that plays an important role in physics. The popular inflation theory, which is hypothesized to explain fundamental features of the early universe, is based on an effective spin-0 particle called the *inflaton*, which has also not been detected experimentally.

WHY IS THE HIGGS SO IMPORTANT?

The Higgs boson is an important player in the particle physics theater, for it is hypothesized that it produces the masses of all the elementary bosons, quarks, and leptons in the standard model. Elementary particles, as we recall, do not contain more basic constituents, in contrast to the protons, neutrons, and mesons. The proton and neutron are not elementary particles because they are composed of quarks, and the mesons, such as the pi meson and K meson, are composed of quarks and antiquarks. These particles belong to a family of composite particles called *hadrons*.

We know from accelerator experiments that the W and Z bosons differ in mass by about 10 GeV, and the photon is known experimentally to have zero mass. These facts have to be explained in the electroweak theory in the standard model of particle physics, as developed first by Weinberg and Salam in 1967/1968. Electroweak symmetry means there was a phase during the early universe when the masses of all the particles except the Higgs boson were zero. The "Higgs mechanism" is invoked in the theory of the standard model to make the spinless scalar field vacuum energy not equal to zero. Normally, the vacuum energy of particles is considered to be zero in modern quantum field theory. Actually, as we know, the vacuum energy is not composed of "nothing," but consists of particles and their antiparticles annihilating one another to produce an energy of zero. The postulated nonzero vacuum energy causes the "spontaneous symmetry breaking" that gives elementary particles their masses.

The Higgs boson is so crucial because it makes the unified theory of electromagnetism and the weak force within the standard model—the so-called *electroweak theory*—renormalizable within modern quantum field theory. This means that we are able to perform calculations of the collisions of particles, for example, that are finite and meaningful, rather than having to deal with impossible infinities. By way of analogy, the early developments of quantum field theory during the 1930s and 1940s produced a renormalizable and therefore finite theory of QED. This theory is one of the most remarkably successful theories in physics. It is the relativistic quantum theory of James Clerk Maxwell's classical electromagnetic field equations. Particle physicists hope that the standard model with the Higgs boson can repeat the success of the renormalizable QED.

WHERE IS THE HIGGS BOSON?

The final results from the Tevatron accelerator in the United States and the results up to March 2012 from the LHC already ruled out great swaths of the mass–energy range possible for the Higgs boson to exist. Only a small window of energy was left in which the Higgs particle could be hiding. As it turns out, this was the most difficult energy window in which the LHC could detect the Higgs—namely, the energy range between 115 GeV and 145 GeV. The LHC was built to investigate particle collisions at very high energies, up to 14 TeV. However, at lower energies, such as between 115 GeV and 145 GeV, the unwanted background or statistical noise is high, making it very difficult to detect the decay products of the Higgs boson and thereby infer its existence. It is in this small, remaining window that the putative signal of the Higgs boson has been found. For the LHC to discover the Higgs boson at these lower energies requires a significant luminosity or intensity of proton–proton collisions, which was reached by July 2012.

By 2012, the top quark and the W boson masses were accurately known experimentally. Armed with this knowledge, one could estimate the mass of the Higgs boson. Moreover, fits to precise electroweak data collected during the past decade or two by the LEP at CERN, the Tevatron collider, and, more recently, LHC experiments, also allowed one to constrain the possible mass of the Higgs boson. Combining all the data shows that the Higgs boson mass

"Always the last place you look!"

Figure 4.1 © Farley Katz, *New Yorker Magazine*

should be light—between 115 GeV and 130 GeV. Indeed, the best fit to the electroweak data puts the Higgs mass at around 97 GeV. However, this value of the Higgs mass was already excluded at LEP, which stopped functioning in the year 2000 to make room for the LHC, which is housed in part in the original LEP underground ring. This exclusion by LEP, however, only has a statistical significance of one standard deviation (sigma), which is not significant enough to exclude a possible Higgs boson from 97 GeV up to an energy of 135 GeV.

By March 2012, spokespeople at the ATLAS and CMS detectors at the LHC were claiming that the proton–proton collision data gathered in 2012 will either rule out the Higgs boson or discover it in the narrow band of remaining, unexplored low energy, which is only 15- to 20-GeV wide. They claimed that this will end the search for the elusive "God particle" that has been going on for more than three decades.

If the discovery of the Higgs boson is not confirmed, this will create a crisis in physics. Indeed, the many textbooks on particle physics and relativistic quantum field theory published during the past 40 years will need to have three or four chapters ripped out. Our whole understanding of the fundamental origin of matter and what makes the unified electroweak theory consistent physically would have to be rethought from scratch.

DO SUPERSYMMETRIC PARTICLES EXIST?

The search for symmetries in nature has been going on since ancient Greece. Symmetry is such a pleasing concept to the human mind that we assume it must be the basis of nature. There are many examples of obvious symmetries, such as symmetric shapes like spheres or snowflakes, or the bilateral symmetry of butterflies or the axial symmetry of trees.

Physicists' search for symmetries in nature led them to a symmetric description of space and time called *supersymmetry*. The use of *super* suggests that this is the most symmetric description of particle physics and spacetime that can be achieved. The idea of supersymmetry was proposed by Hironari Miyazawa in 1966 and was rediscovered subsequently during the early 1970s by several physicists, including Y. Golfand, E. P. Likhtman, Julius Wess, and Bruno Zumino. The origin of supersymmetry lay mainly in mathematical speculations about the possible maximum symmetries of space and time. However, the concept was soon adopted by particle physicists for more practical purposes.

When supersymmetry is translated into particle physics, it requires that the number of particles that are observed in the particle zoo be doubled. Each of these new particles is a superpartner of an existing one; one will be a boson with half-integer spin and the other a fermion with integer spin.[1] That is, the spin of the superpartners always differs by a spin unit of ½. For example, the electron has spin ½, whereas its superpartner, the selectron, has spin 0. Without supersymmetry, only the bosons are the carriers of forces between particles; but, in supersymmetry theory, fermions can also play this role. For example, the photon, a boson and the carrier of the electromagnetic force, has spin 1, whereas its superpartner, the photino, a fermion, has spin ½ and carries a new supersymmetric force that acts between the superpartners.

HOW IS SUPERSYMMETRY USEFUL?

One of the earliest applications of supersymmetry was in attempting to solve the so-called *Higgs mass hierarchy problem*. According to the standard model of particle physics, the mass of the Higgs boson receives an enormous contribution from the interaction of this particle with itself—its so-called *self-interaction*. This is in contrast to the masses of the quarks and leptons, and the W and Z bosons, the masses of which are generated directly by the interactions of the Higgs boson, or Higgs field, with these elementary particles. The Higgs mass

1. Fermions have half-integer spins. For example, the electron has spin ½. Bosons, on the other hand, have integer spins, such as the photon with spin 1.

hierarchy problem arises because, to get a down-to-earth value for this mass, physicists have to perform a tremendous fine-tuning—a very delicate mathematical cancelation involving numbers with a great many decimal places—between the "bare mass" of the Higgs boson, which is the mass in the absence of interactions with other particles, and the contribution coming from the self-interaction. Such fine-tuning in physics produces a very unnatural consequence for theoretical calculations, and it is unacceptable to most theoretical physicists. One of the major challenges in particle physics during the past four decades has been to remove this Higgs mass hierarchy problem. The most efficacious way to do this has been with supersymmetry. This solution has been called *natural supersymmetry.*

Another notorious fine-tuning problem in particle physics is the calculation of the energy density of the vacuum using relativistic quantum field theory. Again, the calculation of the vacuum energy density can produce a startling disagreement, by as much as 10^{122}, with the expected observed value of the vacuum energy density. This absurd result is considered one of the worst predictions in the history of physics. In particular, when the vacuum energy density is calculated from the Higgs field vacuum, it produces by itself this absurd fine-tuning of the vacuum energy density. See Chapter 10 for a more detailed discussion of the concepts of fine-tuning, the Higgs boson mass hierarchy problem, the gauge hierarchy problem, and the cosmological constant problem.

Returning to supersymmetry, Leonard Susskind and others during the early 1970s suggested that if superpartners really existed, then this could solve the Higgs mass hierarchy problem. Technically speaking, this required the masses of the superpartners to be not too different from the known masses of the quarks and leptons. In this interpretation, some suggested that a "supercharge" on the superpartners existed in addition to ordinary electric charge. This interpretation resulted in a cancelation between the supercharges of the particles and their superpartners, which alleviated the Higgs mass hierarchy problem. In other words, during the calculation of the Higgs mass, the positive particle contributions are canceled by the negative superpartner contributions, which makes the mass of the Higgs boson agree with its anticipated experimental value. A similar cancelation in a supersymmetric theory would resolve the serious fine-tuning problem in the calculation of the vacuum energy density of particles.

During the 1970s and 1980s, Peter van Nieuwenhuizen, Sergio Ferrara, and Daniel Freedman, among others, introduced the framework of supergravity, in which the boson force carrier of gravity, named the *graviton*, with spin 2, has a superpartner called the *gravitino* with spin ³⁄₂. The supergravity theory can help solve the problem of how to unify gravity with the other three forces of nature.

Another important problem that supersymmetry may solve is the cosmological constant problem. We recall that the calculation of the vacuum energy density is absurdly big compared with observational data. This vacuum energy density can be related directly to the so-called *cosmological constant*. Einstein introduced the cosmological constant in 1917 into his gravitational field equations to make the cosmological model based on his gravity theory lead to a static universe. This was the first paper introducing modern cosmology based on Einstein's gravity theory.[2] Then, during the 1920s, the Russian cosmologist Alexander Friedmann solved Einstein's field equations and discovered that the universe of general relativity was dynamical and would undergo an expansion. This was later found independently by Belgian cosmologist Georges Lemaître. With astronomer Edwin Hubble's discovery in 1929 that the universe is indeed expanding, the need for Einstein's cosmological constant disappeared. However, it has reappeared in today's so-called *standard model of cosmology* to explain the apparent *acceleration* of the expansion of the universe, which astronomers discovered through supernovae data in 1998. Einstein's cosmological constant, designated by the Greek letter Lambda, Λ, produced a repulsive, antigravity force in the universe that was able to balance the attractive force of gravity, as was originally required by Einstein in his static model of the universe. As it is interpreted today by cosmologists, the vacuum energy density associated with the cosmological constant produces a repulsive force that accelerates the expansion of the universe. The energy associated with this force is called *dark energy*.

During the mid 1960s, Russian physicist Yakov Zeldovich identified the cosmological constant with the energy of the vacuum. The seeming sea of annihilating particles and antiparticles that constitutes the modern vacuum produces a constant vacuum energy that is identified with the cosmological constant through Newton's gravitational constant. Calculating this vacuum energy leads to the enormous discrepancy of more than 120 orders of magnitude between the theory and observation, and constitutes the most extreme fine-tuning problem in all of physics, signaling a major crisis in both cosmology and particle physics.

In supersymmetry, because of the cancelation of energies between particles and their superpartners, this extreme fine-tuning of the vacuum density no longer occurs. However, we now know experimentally, as of March 2013 from LHC results, that no superpartners exist below a mass of 600 to 800 GeV. In addition, the gluino, which is the superpartner of the gluon, has been excluded

2. Albert Einstein, "Kosmologische Betrachtungen zur allgemeinen Relativitätstheorie," *Königlich Preussische Akademie der Wissenschaften*, 142–152 (1917).

up to 1.24 TeV. This promising solution of the cosmological constant problem has been dealt a serious blow.

If the search for superpartners continues to come up empty-handed at the LHC, then it would remove one of the early motivations for promoting supersymmetry. The theory would fail to solve the Higgs mass hierarchy problem and the fine-tuning problem associated with the vacuum energy density calculations. It now appears that if, indeed, superpartners exist, they will have masses beyond what can be detected by the LHC.

SUPERCASTLES IN THE AIR

In its early days, supersymmetry also played an important role in the development of string theory with its many extra dimensions. The original version of string theory contained only boson particles with integer spin. Relativistic quantum field theory and particle physics were founded on the concept that particles are points in space—that is, they had zero dimensions—whereas, in string theory, particles are one-dimensional strings. These strings are very tiny, only about 10^{-33} cm, which is about the Planck length, and it is the frequencies of their vibrations that identify them as different particles. This theory requires a total of 26 dimensions, which is in stark contrast to our known universe of three spatial dimensions and one time dimension. The need for the extra dimensions arose because of the requirement that string theory satisfy Einstein's special relativity, which only has three spatial dimensions and one time dimension in what is called *Minkowski spacetime*.[3] The mathematics of string theory requires that 26 dimensions are necessary to compactify, or shrink, the dimensions to four. It was eventually discovered that supersymmetry, with all its partner particles, was needed to make string theory consistent physically with fermions as well as bosons, which led to what is now called *superstring theory*, whose viability depends on the discovery of superpartners by the LHC and future accelerators.

Superstring theory has declined in popularity during the past few years because it has not been possible for string theorists to propose realistic tests of the many versions of this theory, except for experimental ways to find the extra space dimensions required by string theory. One of the major claims of superstring theory is that it leads to a finite theory of particle physics—that is, that the standard model of particle physics falls out of the equations of string theory. However, this has not been proved satisfactorily. Nor has it been

---

3. The algebra of the operators of string theory for boson strings is only consistent with special relativity in 26 dimensions, thereby satisfying the symmetry of Lorentz invariance.

proved rigorously that replacing zero-dimensional points in spacetime with one-dimensional strings makes finite quantum calculations possible. Because superstring theory is claimed to unify all the forces of nature, including gravity, it would also lead to finite calculations in quantum gravity. However, the latest experimental results from the LHC have excluded the existence of extra dimensions, and weakened the viability of superstring theory, up to about 3 TeV, which represents a lot of territory.

The story about extra dimensions starts with the publication of a unified theory by Theodor Kaluza, born in Silesia in 1885. He was inspired by Einstein's attempt to unify gravity and electromagnetism, the only known forces during the 1920s. Kaluza proposed that there is a fourth spatial dimension in addition to the three known ones, making spacetime five-dimensional when we include the dimension of time. Oskar Klein, born in 1894, developed Kaluza's theory further, and compactified the fifth dimension to a very small size so that we could understand why we had not observed the fifth dimension yet.

To verify this theory, the LHC has been searching for what are called *Kaluza–Klein particles*, which are predicted when one generalizes Einstein's field equations to five dimensions. The Kaluza–Klein particles are proposed to exist only in the fifth dimension. Visualize a city street and a car moving along it in our known four-dimensional spacetime. However, we can also imagine the car veering to the sides of the street, and this extra degree of freedom corresponds to having more than three spatial dimensions. So when we measure the momentum of the car moving along the street without veering to one side or the other, we are measuring the momentum of the car in our spacetime dimensions. If we measure the momentum of the car as it veers to the left or the right on the street, this would correspond to the car's mass or momentum in the fifth dimension.

According to Kaluza–Klein theory, the known particles, such as the electron, would have partner particles with similar physical properties to the electron, but with a mass greater than the mass of the electron (and also greater than the muon and the tau, which are known, observed heavy electrons). The same can be said for the quarks. Are the heavier, known particles such as the muon and the tau Kaluza–Klein partners of the lighter electron? The answer is no. The electron is electrically charged and is surrounded by an electric field. The photon carries the electromagnetic force between electrons and it travels in the same spatial dimensions as the electrons. Therefore, if the electron has Kaluza–Klein partners, then the photon must also have them, and a photon should exist with a mass of about the muon mass. However, such a photon has never been detected. It follows that the muon is not the Kaluza–Klein partner of the electron.

For the Kaluza–Klein particles, if indeed they exist, their extra mass or momentum exists in the fifth dimension. This extra mass can be expressed as Planck's constant h divided by the speed of light c multiplied by length L, which is associated with the extra dimension. For the tiny values of length L resulting from the compactification of the fifth dimension, we have a corresponding mass greater than the measured mass of the electron, and the LHC should be able to detect this extra mass. The magnitude of length L is the compactified length, which is a billionth of a meter or less. Thus, detecting a Kaluza–Klein particle with all the properties of, say, an electron but with a larger mass would verify the existence of an extra dimension. So far, no such Kaluza–Klein particles have been discovered at the LHC up to an energy of more than 2 TeV, at increasingly small compactified lengths. The exclusion of both superpartners and extra dimensions by the LHC has cast a cloud over the whole idea that supersymmetry and superstrings actually exist in nature.

NO DARK OR BLACK THINGS EITHER?

Another exotic prediction of supersymmetry is the existence of dark matter, which is required to explain the dynamics of galaxies, clusters of galaxies, and the standard model of cosmology, also known as the *concordance cosmological model*. There is much stronger gravity "out there" holding together galaxies and clusters of galaxies than is predicted by Einstein and Newtonian gravity, which has made it necessary to postulate extra, invisible matter in the universe to create stronger gravity. This dark matter has so far only been inferred from gravitational phenomena in astrophysics and cosmology. It is based on the idea that Einstein's general relativity is universally correct and does not need any modification. Modified theories of gravity, including my own—called *modified gravity*, or *MOG*—have been published that do not require undetected dark matter and yet can explain the current observational data for galaxy and cluster dynamics and cosmology. However, the majority of physicists and cosmologists believe that dark matter exists and that Einstein's general relativity does not need any modification.

There has been a continuing effort to detect dark matter in underground experiments, at the LHC, and using astrophysical data. There are several candidates for dark-matter particles. The lightest, stable superpartner in supersymmetry, called the *neutralino*, is a popular candidate for the dark-matter particle. Actually, there are four neutralino particles, superpartners of the neutrino, and the candidate for dark matter is the lightest stable partner of this quartet. This hypothetical particle has so far not been detected at the LHC, up to a high energy of about 1 TeV.

This neutralino may be a kind of dark-matter particle called a WIMP (or, weakly interacting massive particle). WIMPs are expected to have a mass of between 2 GeV and 100 GeV, and are classified as belonging to the category of "cold dark matter," an important ingredient in the standard cosmology or concordance model, also called the *LambdaCDM model*. Dark-matter detection experiments are being performed deep underground, to remove the background effects caused by cosmic rays coming in from outer space. Until now, the several large underground experiments and astrophysical observations have not succeeded in finding a WIMP.

Another candidate for dark matter is the axion particle. In contrast to WIMPs, the axion is a very light particle. It can perform the required role of producing dark-matter haloes in galaxies, which could explain the dynamics of galaxies, thereby keeping Newtonian and Einstein gravity unmodified. However experiments for the past two or three decades have not been able to find axions either. All these experimental null results have so far dashed experimentalists' initial hopes of identifying the elusive dark-matter particles.

Theorists with exuberant imaginations dreamed up yet another possible candidate for detection at the LHC: the mini or micro black hole. The idea is that, because regular black holes are formed by the collapse of stars under their own weight, the particles accelerated at the LHC should reach a high enough energy and mass that they, too, could collapse and form mini black holes. Stephen Hawking has gone so far as to suggest that such mini black holes could actually be hiding the Higgs boson. This led him, some years ago, to make a famous bet with Gordon Kane, a professor of theoretical physics at the University of Michigan, that the Higgs boson is either permanently hidden or does not exist. In the event that a mini black hole could be produced at the LHC, some theorists have speculated that it may be possible to examine them and detect the famous "Hawking radiation" supposedly produced at the event horizons of black holes.

Some alarmists feared that if the LHC did produce such mini black holes, they would destroy the nearby city of Geneva and large parts of Switzerland and France, possibly even the whole planet. As we now know, after several years of the successful running of the LHC, Geneva has not been destroyed by mini black holes. In fact, the latest LHC results make it unlikely that mini black holes exist.

WHITHER PHYSICS IF THE LHC CONTINUES TO COME UP EMPTY-HANDED BEYOND THE STANDARD MODEL?

As of early 2013, increasing evidence suggests that the LHC may have discovered the Higgs boson. However, no new physics has been discovered beyond the particles and forces of the standard model. If this situation continues, it raises serious questions about the future of particle physics. The tax-paying public and the governments of the many countries involved in the world's largest experiment will wonder how the theorists could have been so wrong for more than half a century. How could so much money have been spent searching for tiny physical phenomena such as supersymmetric particles and dark matter that do not appear to exist? Despite the possible discovery of the Higgs boson, which would confirm the conventional standard model, does this lack of new physics signal the end of particle physics?

As Guido Altarelli mused after my talk at CERN in 2008, can governments be persuaded to spend ever greater sums of money, amounting to many billions of dollars, on ever larger and higher energy accelerators than the LHC if they suspect that the new machines will also come up with nothing new beyond the Higgs boson? Of course, to put this in perspective, one should realize that the $9 billion spent on an accelerator would not run a contemporary war such as the Afghanistan war for more than five weeks. Rather than killing people, building and operating these large machines has practical and beneficial spinoffs for technology and for training scientists. Thus, even if the accelerators continued to find no new particles, they might still produce significant benefits for society. The Worldwide Web, after all, was invented at CERN.

Not discovering any new physics beyond the Higgs boson at the LHC or future accelerators would have profound significance for the future of physics. Although many physicists consider that possible outcome a looming disaster, others, including myself, prefer to take a contrarian and positive attitude: not finding any new physics at the LHC beyond the standard model would be exciting! Back in 1879 (the year of Albert Einstein's birth) the ether, which was supposed to be the all-pervasive medium and carrier of electromagnetic waves, was not discovered by the experimentalists Michelson and Morley. That important null experiment, and the later ones that confirmed the original results, did not herald the end of physics. Indeed, in 1905, freed from the concept of the ether, Einstein, building on earlier work by Henri Poincaré and Hendrik Lorentz, revolutionized our understanding of space and time with his theory of special relativity. Also, discoveries by Einstein, Planck, de Broglie, and others ushered in the whole quantum revolution in physics.

We can anticipate a similar revolution in physics if we truly discover that there are no new particles beyond those already observed in the standard model. Indeed, many physicists consider the current standard model of particle physics—based on the observed quarks, leptons, W and Z bosons, the gluon, the photon, and now perhaps the Higgs boson—to be unsatisfactory. Why? Because in its basic form, it has 20 free parameters that have to be fitted to data without understanding their physical origins properly. For example, the interaction or "coupling" of the Higgs field to the quarks and leptons does not determine the specific values of their masses, so the coupling constants measuring the strength of the Higgs field to these fermion masses have to be fitted to their observed masses "by hand," rather than by being predicted by the theory. Hence, the fermion masses are free parameters.

Therefore some physicists, including myself, anticipate that there must be new physics beyond the standard model—beyond this mere roster of particles and forces—that could reveal a more unified picture of the particle forces, including gravity. Such a unified framework would then reduce the number of unknown parameters to a very few, even, possibly, zero. The question arises: Can we conceive of a theory that could unify the forces of nature and achieve a very economical form using only the known observed particles? Answering this question could constitute the next revolution in physics.

Discovering the Higgs boson will validate the standard model of particle physics, but it will leave us with serious problems, such as the very unnatural explanation for the Higgs mass itself, unless new physics beyond the standard model is discovered.

5

The Higgs Particle/Field
and Weak Interactions

I joined the newly minted particle physics group at Imperial College London in 1959 as Abdus Salam's first postdoctoral fellow, holding a Department of Scientific and Industrial Research fellowship that paid me £10 a week. Salam had recently been appointed professor of mathematical physics at Imperial College London. He had been brought in as the Cambridge University superstar who would rejuvenate theoretical physics at Imperial College by starting up a group in the very active field of particle physics.

When newly appointed research students to be supervised by Salam asked him what they should pursue as a research project, he would say, "Go and read Julian Schwinger's paper on fundamental particle interactions. The scalar sigma particle may be a clue as to how particles get their mass."[1] Salam had in mind that a renormalizable quantum field theory, one that produces finite calculations of physical quantities, must have a gauge invariance, which implies that the force-carrying particles are massless, as is the case with the massless photon in QED. However, the weak interaction had to be mediated by a massive particle so that the weak force could be short range compared with the infinite-range force of classical electrodynamics mediated by a massless photon. The so-called *sigma particle* represented a scalar, spin-0 field and, through Julian Schwinger's published papers in the 1950s, the ubiquitous scalar field of theoretical physics began to make its appearance. The fact that no experiment had ever detected an elementary spin-0 particle in high-energy accelerators did not deter the theorists from speculating about the consequences of such a particle in particle physics. The experimentally confirmed elementary fermions and bosons in the

---

1. J. Schwinger, "A Theory of the Fundamental Interactions," *Annals of Physics*, 2, 407–434 (1957).

standard model all have spin ½ or spin 1, with the exception of the scalar spin-0 Higgs boson, which was unknown in 1959, and has possibly now been found.

When I first arrived at Imperial College, I made an appointment to see Salam in his large office, which had a rich Persian carpet spread across the floor. Salam sat behind an expansive wooden desk.

"Abdus, do you have any suggestions for what I should pursue in my research?" I asked.

He looked at me intensely with his brown eyes. "John, read Julian Schwinger's paper on the sigma model. Maybe you can explain how particles get their mass."

I sat in the comfortable chair in front of his desk and, averting my eyes, thought about this proposal.

"How particles get their mass?" I asked. "Why is this important?"

Salam grunted and said, "Of course it's important. How are we going to explain weak interactions and beta decay if we don't have gauge invariance in weak interactions, and we have to worry about the bothersome necessary mass of the W particle?"[2]

"So you are saying that this massive W, which has not been found experimentally, doesn't lead to a renormalizable theory, and therefore the weak-interaction calculations become infinite and meaningless. The theory is, in fact, unrenormalizable?"

Salam replied, "Yes. That's it. That's the problem."

"Are you suggesting that this sigma particle—this scalar particle—somehow resolves this problem of infinities? I don't see how this can happen."

Salam frowned and stated, "Well, I don't understand how it happens either, but maybe there's some clue here as to how to resolve the problem."

The problem, of course, was how to give the W boson a mass and yet keep the calculations finite in the theory of weak interactions.

Not long after my arrival at Imperial College, Peter Higgs, an English theoretical physicist, joined the group as a postdoctoral fellow and shared my smartly furnished office near Salam's office. I presumed at the time that he also asked Salam what he should work on as a research project, and I surmised that Salam had said, "Peter, go away and read Julian Schwinger's paper on the sigma model. Maybe it's got something to do with how the particles get their mass."

---

2. The range of a force in particle physics is determined by the mass of the particle. For classical electromagnetism, the fact that the photon is massless results in the electromagnetic force having infinite range. On the other hand, in weak interactions, which govern the radioactive decay of particles, the force has to be short-range, so we know that the intermediate vector boson, W, that carries the weak force has to have a mass. The "confined" massless gluon is an exception.

In the meantime, I read Schwinger's paper and wasn't excited by it, and didn't see how it could resolve the problem of getting rid of infinities in the weak-interaction calculations. I could see how the self-coupling of this scalar sigma particle, or the interaction of the particle with itself, could give mass to other particles—by the coupling of the scalar sigma field with the W boson—but it didn't strike me as particularly convincing, and I moved on to other research.

BANISHING INFINITIES

The tremendous success of the theory of QED of photons and electrons achieved by Feynman, Tomanaga, and Schwinger in 1949/1950 owes much to their development of renormalizable quantum field theory. In this theory, the meaningless mathematical infinities that occur in the calculation of scattering amplitudes could be canceled out by the "bare" mass of the electron—that is, the mass of the noninteracting electron—and with the effective self-mass caused by the interaction of the electron with its own electromagnetic field. The bare mass and the self-interaction mass of the electron are both infinite in the calculations, but these infinities cancel one another, resulting in the finite observed mass of the electron. Paradoxically, although the initial calculations of scattering amplitudes were infinite and meaningless, the final, resulting scattering amplitudes using renormalizable quantum field theory involving interactions of electrons and photons were finite and independent of any cutoff in the energy that was used in the calculations of these amplitudes. Physicists chose this cutoff of infinite energies arbitrarily to make the provisional calculations finite. However, the final calculation after renormalization has been executed must not depend physically on such an arbitrarily chosen cutoff energy. Freeman Dyson demonstrated that a renormalizable theory is, in fact, finally independent of the arbitrary cutoff energy in his famous papers published in 1949 and 1952.[3] As we recall, however, Paul Dirac, who was one of the main protagonists in the invention of QED, never accepted this renormalization scheme.

Particle physics as theorists practice it today relies heavily on the more modern developments of renormalization theory, which is also used in investigations in condensed-matter physics. These developments play an important role in the formulation of QED. The success of QED depends primarily on the photon being massless, which allows the theory to be renormalizable. This fact

3. F.J. Dyson, "The Radiation Theories of Tomonaga, Schwinger and Feynman," *Physical Review*, 75, 486–502 (1949); and   "Divergences of Perturbation Theory in Quantum Electrodynamics," *Physical Review*, 85, 631–632 (1952).

allows one to prove that the QED based on the quantization of Maxwell's classical field equations is gauge invariant. The feature of gauge invariance is crucial for the physical consistency of the theory. For example, it leads directly to the conservation of the electric charge current of the electron and matter fields.

The gauge invariance of QED gives rise to a set of mathematical identities discovered by John Ward during the early 1950s, called the *Ward identities*. These identities guarantee that the probabilities of scattering of photons and electrons never add up to more than 100 percent. In the initial version of Yang and Mills's nonabelian gauge theory extension of Maxwell's equations, the charged vector field was massless, which guaranteed that the theory had a gauge invariance, and the theory could be made renormalizable. It also satisfied the so-called *unitarity condition for the S-matrix*, guaranteeing that the probability of the scattering of particles never exceeded 100 percent. However, we recall Pauli's objection to the original Yang–Mills assumption that massless charged vector mesons exist in nature. As we know, such particles do not exist in nature.

A serious problem arises immediately with the Yang–Mills theory and renormalizability: not all force-carrying bosons are massless! What does this do for the essential feature of gauge invariance in quantum field theory? The immediate answer is that it destroys the gauge invariance, and therefore the theory is no longer renormalizable and no longer conserves scattering probabilities. This problem was recognized already by Yang and Mills, and has been a topic of research since the 1950s, starting with Schwinger and Lee and Yang, and continues to be pursued by many physicists up to the present day. One possible resolution of this problem has been the development of the standard-model electroweak interactions involving spontaneous symmetry breaking and a Higgs particle. The idea of a Higgs boson whose lowest energy state breaks spontaneously the basic group symmetry of SU(2) × U(1) and allows for different masses of the W and Z bosons to emerge within the standard-model electroweak theory, while keeping the photon massless, has been the most popular resolution of this problem.

Fortunately, QCD, the theory of the *strong* interactions of particles, developed in 1973 by Gell-Mann, Fritzsch, and Leutwyler, using nonabelian SU(3) quantum field theory with eight colored gluons, does not suffer from the consequences of having massive force-carrying bosons because the eight colored gluons are massless. It can be demonstrated that this strong-interaction theory has an extended form of gauge invariance, and the Ward identities in it have been generalized by Yasushi Takahashi, Andrei A. Slavnov, and John C. Taylor. The technical issues involved in proving the renormalizability of the nonabelian gauge theory for strong interactions are somewhat formidable. Feynman, in his research on quantum gravity, which is a form of nonabelian gauge theory, with gravitons being the massless force carriers of gravity, discovered that certain "ghost" fields had to be included in the calculations of QCD to guarantee renormalizability. Russian theoretical physicists Ludvig Faddeev and Victor Popov played an important role in developing the mathematics of these ghost fields, which actually do not appear as physical particles in calculations of the scattering amplitudes. A new gauge symmetry was discovered by Carlo Maria Becchi, Alain Rouet, and Raymond Stora, and independently by Igor Tyutin (called *BRST*) that clarifies the deeper meaning of how these ghost fields enter into the renormalizable nonabelian calculations.

The problem of the impossibility of having a renormalizable quantum field theory of *weak* interactions became critical during the late 1950s and early 1960s. A massive intermediate charged vector boson, W, which was necessary to make the weak interaction a short-range force, destroyed any possibility of having a renormalizable and finite theory of weak interactions. Because the W was massive, the basic gauge invariance of the theory was lost. In a seminal paper published by Sheldon Glashow in 1961,[4] in which he introduced the need for electrically neutral currents in weak interactions, mediated by the exchange of a neutral vector particle called Z, he put the masses of the W and the Z "by hand" into the weak-interaction model. He was aware that this would ruin the possibility of getting finite scattering amplitudes involving the W and the Z particles, and the leptons such as electrons and muons. If one entertained the idea that weak interactions began in some phase in the early universe with massless particles, then this initial phase of the theory could be gauge invariant, renormalizable, and finite, just like QED with the massless photon. But, then, where would the masses of the W and the Z and the fermions come from?

SOLID-STATE PHYSICS TO PARTICLE PHYSICS

During the early 1960s, more theoretical physicists began to speculate on the nature of symmetry breaking in quantum theory. Previously, Werner Heisenberg in Germany, among others, had found an interesting new way of explaining ferromagnetism. He viewed atoms in a metal, such as iron, as little bar magnets with a north pole and a south pole. When the temperature of the metal was above a critical point called the *Curie temperature*, the little bar magnets were oriented randomly and the system was rotationally invariant under rotations of the group O(3). Below the critical Curie temperature, the little bar magnets aligned themselves in a certain direction, and the rotational symmetry of O(3) was broken spontaneously.

4. S.L. Glashow, "Partial-symmetries of Weak Interactions," *Nuclear Physics*, 22, 579–588 (1961).

Jeffrey Goldstone, a theorist who was once a colleague of mine at Trinity College Cambridge, theorized that this phenomenon of spontaneous symmetry breaking in solid-state physics could be imported into relativistic quantum field theory. He published a paper in 1961[5] in which he demonstrated that, within a certain quantum field theory model involving a scalar spin-0 field like Schwinger's sigma field, you could invoke spontaneous symmetry breaking of a mathematical group. This symmetry breaking always predicted a massless, scalar spin-0 particle. A similar result had been obtained by Yoichiro Nambu in 1960, and by Nambu and Giovanni Jonah-Lasinio in 1961, when they investigated the nature of superconductors, which are metals cooled to such a low temperature that the electrical resistance of the electrons moving through the metal approaches zero, meaning that the electrons move freely through the metal. Nambu discovered that if the superconducting nature of the material was caused by spontaneous symmetry breaking of a symmetry at higher temperatures, then this predicted the existence of a massless spin-0 particle. These ideas became known as *Goldstone's theorem* or, later, as the *Nambu-Goldstone theorem*.

However, if you invoke this phenomenon in quantum field theory and try to relate it to reality, then you run into the problem of predicting the existence of a massless particle that had never been observed. The only massless particles known at the time were the photon, which has spin 1, not spin 0, and the neutrino, with spin ½. This situation led to an impasse in theoretical particle physics that lasted about three years. Salam and Weinberg, who met with Goldstone when he visited Harvard, investigated this problem and tried to resolve it, but were unable to do so successfully.[6] Another theorist, my former colleague at Trinity College Cambridge, Walter Gilbert, published a paper claiming that spontaneous symmetry breaking in particle physics could not be considered physical.[7] With this impasse, it was not possible even to begin to attempt to explain how the W and Z bosons in weak interactions got their masses, because any such mechanism would be accompanied by these nonexistent, massless scalar particles.

Meanwhile, back in solid-state physics, Philip Anderson, a theorist at Princeton, discovered that when you invoke spontaneous symmetry breaking

5. J. Goldstone, "Field Theories with Superconductor Solutions," *Nuovo Cimento*, 19, 154–164 (1961).

6. J. Goldstone, A. Salam, and S. Weinberg, "Broken Symmetries," *Physical Review*, 127, 965–970 (1962).

7. W. Gilbert, "Broken Symmetries and Massless Particles," *Physical Review Letters*, 12, 713–714 (1964).

to explain superconductivity in materials, the photon becomes massive. The spontaneous symmetry-breaking mechanism had generated an "effective mass" for the photon without producing the pesky massless scalar particles suggested by the Nambu-Goldstone theorem. How had this come about?

In 1963, Anderson published a paper[8] explaining that the reason this happened was because the spontaneous breaking of the U(1) abelian symmetry occurred in such a way that the massless photon field "ate" the Nambu-Goldstone scalar field boson, thereby putting on weight and acquiring mass. In Anderson's paper, he refers to a seminal paper titled "Gauge Invariance and Mass," published by Julian Schwinger in *Physical Review* in 1962,[9] in which Schwinger discusses the significance of gauge invariance and the origin of masses of the elementary particles. It is noteworthy that at the end of Anderson's paper, he discusses relativistic Yang–Mills gauge bosons and the Goldstone bosons in a relativistic context. He ends the article with the prophetic statement, "We conclude, then, that the Goldstone zero-mass difficulty is not a serious one, because we can probably cancel it off against an equal Yang–Mills zero-mass problem." We see that Anderson uses the language of relativistic field theory. This indicates that Anderson comprehended the importance of a so-called Higgs mechanism a year before the publication of the papers in *Physical Review Letters* by Brout and Englert,[10] Higgs,[11] Guralnik, Hagen, and Kibble[12] in 1964. Therefore, it is perhaps justified to include Philip Anderson's name among those who discovered the Higgs mechanism. However, Anderson's work was in the context of nonrelativistic solid-state physics. Particle physicists generally ignored solid-state physics. In fact, Gell-Mann was known to call it "squalid-state physics."

How physicists realized that the photon inside the superconductor actually became a heavy photon is a long story. The phenomenon of superconductivity was first discovered in 1908 by Dutch physicist Heike Kamerlingh Onnes. In contrast to normal electrical conducting materials, which have some

8. P.W. Anderson, "Plasmons, Gauge Invariance and Mass," *Physical Review*, 130, 439–442 (1963).

9. J. Schwinger, "Gauge Invariance and Mass," *Physical Review*, 125, 397–398 (1962).

10. F. Englert and R. Brout, "Broken Symmetry and the Mass of Gauge Vector Mesons," *Physical Review Letters*, 13, 321–323 (1964).

11. P.W. Higgs, "Broken Symmetries and the Masses of Gauge Bosons," *Physical Review Letters*, 13, 508–509 (1964).

12. G.S. Guralnik, C.R. Hagen, and T.W.B. Kibble, "Global Conservation Laws and Massless Particles," *Physical Review Letters*, 13, 585–587 (1964).

resistance to electrical currents, when "superconductors" are cooled close to 0 K—for example, liquid nitrogen at temperatures of –321°F, or 77 K—they have no resistance to electrical currents. At very low temperatures, the material reaches a critical temperature, which varies with each superconducting material, at which the electrons can travel through the material without losing heat or energy. The atoms in these supercooled materials assume the configuration of a lattice, with the electrons moving freely among the atoms within the lattice. This is how metals conduct heat and electricity. As the metal cools down, the repulsion between two electrons is reduced, and they actually bind together to form what is called *Cooper pairs*, named after theoretical physicist Leon Cooper, who did a theoretical investigation of superconductivity. In 1957, John Bardeen, Cooper, and John Robert Schrieffer published a paper[13] in which they explained the origin of superconductivity in metals by means of the Cooper pairs of electrons. It is the Cooper pairs, which form a composite "condensate" of pairs of electrons, that enable the electrons to move freely through the metal without resistance.

Another significant physical phenomenon associated with superconductors is the *Meissner effect*, which was discovered by German physicists Walther Meissner and Robert Ochsenfeld[14] in 1933 by measuring the magnetic field outside superconducting tin and lead samples. They discovered that the superconducting metals expelled the magnetic fields—actually pushed the fields away in space. These experiments demonstrated that superconductors have a unique physical property beyond reducing electrical resistance to zero.

The history of the theory of superconductors is a long one. Original work done by Vitaly Ginzberg and Lev Landau, first published in 1950, seven years before the publication of the paper by Bardeen, Cooper, and Schrieffer, provided an effective theory of how superconductors work.[15] In the Ginzberg–Landau theory, an "order parameter" measured the degree of transition the metal attained when it went through a critical temperature threshold, giving rise to what is called a *phase transition*. When the transition temperature was reached, and the system was at its lowest energy state—the ground state—then electron (Cooper) condensates formed and a superfluid was created. The Ginzberg–Landau phenomenon was described by a wave function in the

13. J. Bardeen, L.N. Cooper, and J.R. Schrieffer, "Theory of Superconductivity," *Physical Review*, 108, 1175–1204 (1957).

14. W. Meissner and R. Ochsenfeld, "Ein neuer Effekt bei Eintritt der Supraleitfähigkeit," *Naturwissenschaften*, 21 (44), 787–788 (1933).

15. V.L. Ginzburg and L.D. Landau, *Zhurnal Eksperimental'noi i Teoreticheskoi Fiziki*, 20, 1064 (1950). English translation in: L.D. Landau, *Collected Papers* (Oxford: Pergamon Press, 1965), 546.

presence of an electromagnetic field. The wave function has the characteristic features associated with the gauge invariance of Maxwell's equations for the electromagnetic field.

Now the Meissner effect comes into play. At a certain length from the surface of the superconductor, called the *screening length*, an external magnetic field cannot penetrate the metal, and the interior magnetic field is expelled macroscopically from the interior of the superconductor. This effect can be explained in terms of an "effective" nonzero photon mass. Indeed, there is a connection between the screening length and the effective mass of the photon, which carries the electromagnetic force. In certain physical units, the mass of the photon is inversely proportional to the screening length; when the photon mass increases, the screening length gets smaller. The physical origin of the screening length can be explained by observing what happens when a magnetic field is applied to a field of charged particles. The magnetic field accelerates the charged particles, resulting in currents that tend to cancel or screen the applied magnetic field.

This is the Meissner effect; it pushes out the magnetic flux from the interior of the superconductor (Figure 5.1). Screening currents are set up within the superconductor over distances about as long as the screening length from the

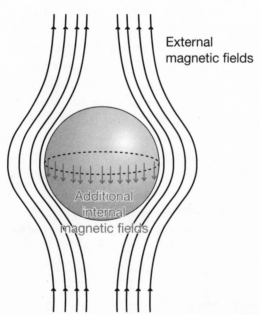

Figure 5.1 The Meissner effect. This diagram shows how the magnetic field is screened as a result of the Meissner effect. © Dwi Prananto, Department of Physics, Tohoku University.
SOURCE: Simpliphy.wordpress.com

exterior boundary of the material. These currents cancel exactly the applied magnetic flux in the interior of the superconductor, and the size of the screening length can be measured to be approximately 10^{-8} of a meter.

The Meissner effect illustrates much of the essential physics involved in the generation of a massive photon inside a superconductor. The Ginzberg–Landau wave field plays an important role in addition to the electromagnetic field. It explains the "discontinuous" transition between the massless photon and the massive one—namely, going from two degrees of freedom to three degrees of freedom. The degrees of freedom are the two transverse and one longitudinal degree of freedom. A massless photon is described by electromagnetic waves that propagate only in transverse directions along the photon's axis of propagation. A massive photon would have an additional degree of freedom—namely, the longitudinal one directed along its axis of propagation. We are now close to understanding why the spontaneous symmetry-breaking Higgs mechanism was "borrowed" from the condensed-matter physics of superconductors.

INVENTING THE HIGGS BOSON IN THE STANDARD MODEL

In 1964, six physicists considered whether the phenomenon of superconductivity and ferromagnetism in solid-state physics could apply to relativistic particle physics. They were still trying to resolve the problem of the massless Goldstone boson, and how to give mass to the vector bosons of weak interactions, the W and the Z particles. Robert Brout and Françoise Englert in Belgium, Peter Higgs in the United Kingdom, and Carl Hagen, Gerald Guralnik, and Tom Kibble also in the United Kingdom considered the interaction of a massless photon gauge field, studied by Philip Anderson in superconducting materials, with a scalar field in the context of a relativistic field theory. Through the interaction of this particle physics gauge field with the scalar field, the irritating massless Goldstone bosons were "eaten" by the vector gauge particle, producing a third degree of freedom, which allowed the gauge field particle to become massive. This compares directly with the phenomena of the superconductors and the Meissner effect, in which the massless photon acquires an effective mass when it interacts with the matter field or Ginzberg–Landau wave function.

Peter Higgs submitted a paper to a journal in Europe, and the editor of the journal at CERN rejected the paper because he did not consider it was of physical significance for particle physics; it was just a mathematical speculation. In his paper, Higgs had offered a simple, elegant model in which a photon with its U(1) electromagnetic gauge field interacted with a scalar field. This broke spontaneously the gauge symmetry of the vacuum associated with this physical system, generating a mass for the U(1) gauge field—that is, the photon.

After his paper was rejected, Higgs included an addendum to his manuscript proposing that the mechanism he discovered required the existence of a new particle, with spin 0, associated with a scalar field. Now the reviewing editor of *Physics Letters* saw the physical significance of this work, for accelerators could hunt for this predicted particle.[16] The other five physicists discussed the same spontaneous breaking of gauge invariance as did Higgs; however, they suggested only indirectly that a new particle had to accompany the mechanism to produce masses for the gauge bosons. Then Higgs, as well as Brout, and Englert, and the trio Hagen, Guralnik, and Kibble published letters in *Physical Review Letters* in the United States within weeks of one another. (See citations in footnotes 10–12.) The letter by Hagen, Guralnik, and Kibble did cite the already published papers of Brout and Englert, and of Higgs. In recent interviews, Guralnik claimed that it was a mistake that he and his collaborators had cited the other papers because it implied that they had come to the subject later, whereas in fact they had come up with these ideas independently and only learned about the other papers after they had drafted theirs. Because Higgs actually predicted the existence of a particle that was responsible for providing gauge bosons with mass, this predicted boson subsequently became known as the Higgs particle (boson), and only recently has the term *Higgs mechanism* been replaced by the incredibly cumbersome but more historically accurate term, the *Brout–Englert–Higgs–Hagen–Guralnik–Kibble* (BEHHGK; pronounced "beck") *mechanism*. To do justice to the seminal published paper by Anderson, this mechanism is even sometimes dubbed the ABEHHGK mechanism!

In 1966, Higgs was visiting the States and was invited to Harvard to give a talk on his mechanism for producing mass for gauge particles through spontaneous symmetry breaking. It appears that some members of the audience, including Sydney Coleman, dismissed the whole idea. However, Steven Weinberg, who was in the audience, was at the time engaged actively in trying to understand how to unify electromagnetism and the weak interactions, and fit in the W and Z bosons of Glashow's 1961 model. It occurred to him, apparently while driving to his Harvard office, as he later explained, that he could incorporate the spontaneous symmetry-breaking mechanism explained in Higgs's lecture into a unification scheme of the weak force and electromagnetism involving leptons. He came up with a model in which the nonzero vacuum expectation value of a scalar field interacting with leptons was able to break spontaneously the gauge

16. P. Higgs, "Broken Symmetries, Massless Particles and Gauge Fields," *Physics Letters*, 12, 132–133 (1964).

symmetry proposed originally by Glashow in such a way that the W and Z bosons would acquire a mass whereas the photon remained massless.

In this initial attempt at an electroweak model, Weinberg considered only leptons such as the electron and muon, ignoring the quarks, to make his model easier to create. He also incorporated into the scheme an angle that allowed him to rotate the Z boson into the photon and vice versa. This meant that a neutral vector boson took upon itself the guise of both a massless photon and a massive Z particle. For a given value of this angle, Weinberg was able to come up with an approximate prediction for the masses of the W and Z bosons, which allowed him to obtain a result that agreed experimentally with the ratio of the neutral current to the charged weak current that had been observed at CERN. Weinberg's seminal paper on this topic[17] provided the basics for what we now call the standard model of particle physics.

A key idea of Weinberg's was to make the Higgs field an isospin doublet expressed in complex numbers—meaning that, in the isospin space, there is a charged and a neutral component to the field. When this isospin doublet interacted with the gauge fields such as the W and the Z, and the quarks and leptons, it allowed—through spontaneous symmetry breaking—the elementary particles to get their masses.

Another important ingredient in developing the standard model was the choice of the potential energy for the scalar field, which was incorporated into the Lagrangian or action principle. This potential was such that it had a nonzero value when the scalar Higgs field vanished. This value of the potential is what is called the *false vacuum,* which is an unstable value. Imagine a Mexican hat (Figure 5.2). A ball rolls down from the top of the hat, or the false vacuum, to a nonzero value of the Higgs field at the bottom of the hat, where the potential energy of the Higgs field is zero. This is the *true vacuum.* If you picture yourself sitting at the false vacuum, or the top of the Mexican hat, and then rotating the hat around the vertical axis, from your point of view the hat keeps its symmetric shape. On the other hand, if you picture yourself sitting at the bottom of the hat brim—at the energy ground state where the scalar field is a constant nonzero value—then, when you look around you at the rotating hat, it no longer retains a rotational symmetry. By moving from the false vacuum to the true vacuum, you have broken spontaneously the symmetry of the group associated with the Mexican hat potential.

Physicists choose the form of the potential energy for the scalar field in a rather ad hoc way. To guarantee renormalizability of the theory, the self-coupling of the scalar field in the potential energy must be of the fourth power of the scalar

17. S. Weinberg, "A Model of Leptons," *Physical Review Letters*, 19, 1264 (1967).

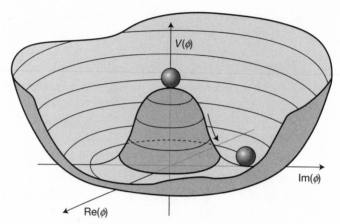

Figure 5.2 Mexican hat potential with the ball initially at the false vacuum.
SOURCE: Luis Álvarez-Gaumé and John Ellis, "Eyes on a Prize Particle," *Nature Physics* (Nature Publishing Group, Dec. 21, 2010).

field with an unknown coupling constant, lambda (λ), determining the strength of the interaction. If the power of the scalar field self-coupling is greater than four, then the theory is not renormalizable. The Mexican hat picture describes, in a simple, generic way, the Higgs mechanism of spontaneous symmetry breaking, provided that the special form of the potential energy for the self-coupling of the scalar field is chosen.

At the same time that Weinberg was working on incorporating spontaneous symmetry breaking into the electroweak gauge theory, Abdus Salam, at both Imperial College and at his International Center for Theoretical Physics in Trieste, was also working independently on the same problem using the same techniques. In a set of unpublished lectures to graduate students, which he gave in the fall of 1967, Salam developed a model very similar to Weinberg's. Unfortunately, there is no written record of these lectures. At a symposium held in Gothenburg in May 1968 on particle physics, sponsored by the Nobel Foundation, Salam gave a talk that he claimed was based on his lectures at Imperial College the previous fall. This lecture, which developed a model involving only leptons, akin to Weinberg's published model of the previous year, was published by the Nobel Symposium in a monograph with a limited circulation.[18] In Glashow's 1961 paper on weak interactions, he had predicted that neutrinos and electrons could scatter off other particles without the exchange of charge through a neutral weak current. This current was associated with a

18. *Elementary Particle Theory: Relativistic Groups and Analyticity*, ed. Nils Svartholm. (Stockholm: Almqvist & Wiksell, 1968).

neutral vector boson, Z. In 1967, Weinberg used Glashow's theory to predict the mass of the neutral Z boson. He then predicted the parity violation of polarized electron scattering, which was observed at Stanford in 1979. In Salam's talk at the Nobel Symposium, he did not give the relationship between the W and the Z masses predicted by Weinberg. Therefore, he was not able to determine the difference in strengths between the weak, charged, and neutral currents. Glashow, Weinberg, and Salam were awarded the Nobel Prize in 1979 for their work on developing electroweak unification.

Eventually the Glashow–Weinberg–Salam model, as it came to be called, was extended to incorporate quarks, making it a more complete theory that began to take on the appearance of what we now know as the standard model of particle physics. This electroweak theory was incorporated into the QCD theory of Gell-Mann, Fritzsch, and Leutwyler, and thus the strong interactions involving the massless colored gluon force carriers of QCD, as well as the massive W and Z bosons and the massless photon of the electroweak theory, were all accounted for in one scheme.

From this historical account, we can see that the idea of spontaneous symmetry breaking of the vacuum in gauge theories in particle physics came directly from the theory of superconductors. From the citations in the published papers by the "Group of Six" in 1964, it is not sufficiently clear to what extent they owe an indebtedness to the pioneers of superconductor physics such as Ginzberg, Landau, and Anderson. However, Higgs did refer to Anderson's paper in a subsequent paper published in 1966.[19] This leads us to the fundamental question: Does nature in its laws for particle physics and quantum fields simply copy what happens in condensed-matter physics? The answer to this question, as we will see as this book continues, relies significantly on the discovery of the Higgs particle. Strictly speaking, however, the answer to the question is no if we identify the scalar Higgs particle as a fundamental, elementary particle, for no such particle exists in the theory of superconductors. The basic entity that plays the role of spontaneously breaking the symmetry of the superconductor is the electron condensate formed from the Cooper pair bound state, which is not an elementary particle. Therefore, it is not clear that carrying through the ideas of the nonrelativistic physics of superconductors to relativistic particle physics will result in the existence of a physical *elementary* scalar Higgs boson.

Both Weinberg and Salam suggested that their electroweak theory was renormalizable; however, they did not provide any proof of this. Martinus Veltman in Utrecht had a brilliant graduate student, Gerard 't Hooft. He proposed to

---

19. P.W. Higgs, "Spontaneous Symmetry Breakdown without Massless Bosons," *Physical Review*, 145, 1156–1163 (1966).

't Hooft that he try to prove that the electroweak theory of Weinberg and Salam was renormalizable. One day, 't Hooft told Veltman that he was actually able to prove that the theory was renormalizable. He had done it! Veltman was astonished and delighted, and in 1971 at a conference he announced how his graduate student, 't Hooft, had proved the renormalizability of the theory. Veltman had contributed significantly to the understanding of weak interactions, and he and 't Hooft collaborated to fill in many of the technical details needed to prove that the claim of renormalizability was robust. Indeed, Veltman wrote a computer program called SCHOONSHIP, which was able to derive the many Feynman diagrams necessary to complete the proof. 't Hooft and Veltman were awarded the Nobel Prize for their contributions to electroweak theory in 1999.

FINDING THE HIGGS BOSON

By the 1970s and 1980s, accelerators began accumulating data that could validate the predictions of the standard model. In particular, confirmations of QCD—such as the verification of colored gluons—convinced most physicists that the standard model was here to stay. More data were accumulated for weak-interaction processes at high energies. The expected bottom and top quarks were discovered at Fermilab's Tevatron in 1977 and 1995, respectively, completing the predicted three generations of quarks. By the late 1990s, all the basic elementary particles of the standard model had been proved to exist, except for the Higgs boson, and they all had spin ½ or spin 1. The standard model now included QCD and the electroweak theory, which in turn contained QED.

Despite the successes of the standard model, confirmed by accelerator results, one essential building block of the whole edifice was missing. Where was the spin-0 Higgs particle? Unfortunately, in contrast to the W and Z boson masses, the mass of the Higgs boson is not predicted by the standard model, which means that within certain theoretical limits, the experimentalists did not know at what energy level to look for it. However, by 2000, when the LEP closed down at CERN, the existence of the Higgs boson had been excluded up to an energy of 114.5 GeV. Moreover, if the Higgs boson mass was less than 120 GeV but above 114 GeV, then the vacuum state associated with the spontaneous symmetry-breaking mechanism would be unstable. This means that, in the worst circumstances, the universe would not last the 14 billion years that was now built into the standard model of cosmology. Moreover, if the Higgs mass was greater than 800 GeV, then the standard model would break down. At a mass of more than 800 GeV, it would no longer be possible to do precise calculations in the standard model using perturbation theory, in which ever-increasing orders of calculations were under control. Also, the theory would violate unitarity and we would start getting probabilities for scattering experiments that exceeded 100 percent, which is impossible. In 2000, this left the energy range from 114.5 to 800 GeV open to experimental detection of the Higgs particle, albeit taking into account the potential problems with instability at less than 120 GeV.

The LHC began operating in 2009, after repairs. In March 2012, the CMS and ATLAS detectors at the LHC had excluded a Higgs particle between 130 GeV and 600 GeV to a confidence level of 95 percent. The fits to the precise electroweak data gathered independently by groups at CERN, the Tevatron, and other theoretical groups, had restricted the Higgs particle to a mass between 97 GeV and 135 GeV. Indeed, the best fit to the precise electroweak data for the Higgs mass was a mass of about 97 GeV, which had already been excluded by the LEP bound of 114.5 GeV. However, in statistical terms, the best fit of 97 GeV was one standard deviation away from the central value of the fit, allowing for a possible Higgs mass between 114.5 GeV and 135 GeV. By March 2012, this was the small window of energy–mass left open in which to detect the Higgs particle, and by late 2012, experimentalists claimed to have detected their quarry at about 125 GeV.

Much like the ill-conceived ether of the 19th century, the Higgs field is supposed to pervade all of spacetime at all times. This Higgs field is believed to interact with the elementary quarks, leptons, and the W and Z bosons, as well as with itself, but not with the photon or gluon. This interaction would produce masses for all the elementary particles except for the photon and gluon. It is as if the particles are swimming through a tank of water that resists their forward motion. The heaviest particles, such as the top quark, would feel the resistance more than the lighter ones, such as the electron and the lighter up and down quarks. The photon and gluon would swim through the water without any resistance at all, because they are massless. The Higgs field/ether would be produced by a phase transition (like steam turning to water when cooled) about 10^{-12} seconds after the Big Bang. This hypothesis is basic to the standard-model electroweak theory.

Despite the fact that, after almost 50 years, the elusive Higgs boson had not been detected conclusively up until March 2012, the idea that it really existed was so entrenched in the minds of particle physicists that almost the whole community expected it to be discovered soon. Why is this so?

One reason is that the Higgs boson could explain the origin of the masses of the elementary particles. However, it does fall short of predicting the specific experimental masses of the fermions, such as the leptons and quarks, for each of these particle masses has a coupling constant associated with it, which determines the strength of the interaction of the particle and the Higgs field. Physicists simply adjust these coupling constants to yield the experimental values of the particle masses; the masses are not predicted by the theory. On the other hand, Weinberg was able to predict the approximate masses of the W and Z bosons given the experimental value of the so-called *weak angle*, which allowed for the rotation of the photon into the neutral Z boson. The Higgs mechanism seemed, to particle physicists, to be the best game in town for explaining the origin of mass for the elementary particles. An even stronger reason for believing in the existence of the Higgs boson is its ability to produce a renormalizable or finite theory of weak interactions that does not violate unitarity.

However, there have been serious problems in believing in the existence of the Higgs boson. As Weinberg and Salam were developing the spontaneous symmetry-breaking mechanism, there was an ad hoc aspect to the scenario that led some theorists to question whether the model was correct. For one thing, they had put the masses and coupling constants of the quarks and leptons in by hand, rather than the theory predicting them. As we recall, another ad hoc feature of the theory was that the self-interaction of the scalar Higgs field was chosen to be a very specific value (the scalar field to the power of four) so that it generated a renormalizable electroweak theory. Other choices that would seem equally plausible would not allow the theory to be renormalizable or to lead to finite calculations.

Another technical issue is that for a scalar spin-0 field like the Higgs field, the divergences occurring in the quantum theory calculation of the Higgs boson mass were quadratic in the energy cutoff, as opposed to the logarithmic divergences for the W and Z particles and fermions. This leads to a serious fine-tuning problem that had to be confronted. The fact that when quantum corrections to the calculation of the Higgs boson mass were performed, which are required when you renormalize the Higgs mass, these quantum corrections were not actually "corrections." Indeed, they were so enormous that they produced a critical fine-tuning disaster for estimates of the Higgs mass. This became known as the "*Higgs mass hierarchy problem*," which we encountered in Chapter 4. This means that the calculation of the Higgs mass would produce absurdly large results compared with what is anticipated to be experimentally valid.

Another serious fine-tuning problem with the Higgs particle, which Martinus Veltman has emphasized for years, is that when one calculates the energy density of the vacuum in the presence of the Higgs ether, this vacuum density could be 56 orders of magnitude larger than has been determined observationally in cosmology. It could even be as much as 122 orders of magnitude larger, when the Planck mass associated with gravity is accounted for. Such a vacuum density would make it impossible for our universe to exist at all. This belongs to the category of what is called the *cosmological constant fine-tuning problem*. Other causes of such a huge vacuum density occur in particle physics, but the one associated with the Higgs boson has made the whole idea of the Higgs mechanism quite unattractive to some physicists.

In summary, although the Higgs boson is essential for the electroweak theory to work, it creates serious problems. Therefore, some physicists, myself included, began to explore alternatives to both the Higgs mechanism and the conventional electroweak theory. Of course, if the evidence of *something* new at 125 GeV turns out definitely to be the Higgs boson, then physicists will have to roll up their sleeves and find solutions to the problems that the Higgs brings with it.

6

Data That Go Bump in the Night

The annual Topical International Conference on Particle Physics, Cosmology, and Astrophysics is always known as the "Miami" conference, even though several years ago it moved from Miami to Ft. Lauderdale, Florida. My talk at "Miami 2011" was scheduled for 11:00 a.m. on December 15, 2011. Its title was "If the Higgs Particle Does Not Exist, Then What?" At the start of the session, the conference organizer, Thomas Curtwright, put my PowerPoint presentation into the conference computer. As my title flashed on the large screen at the front of the auditorium, Thomas laughed and said, "Well, in view of Tuesday's press conference at CERN, I guess your talk has crashed."

I frowned at him. "Don't be so fast with the conclusions. I'll explain in my talk why the experimental results are still inconclusive."

The auditorium in the Lago Mar resort where the conference was being held was filled with physicists. In the audience was Lars Brink, who is currently the only particle theorist on the Stockholm Nobel committee. Also in the audience was Françoise Englert, one of the Group of Six who proposed the mechanism of spontaneous symmetry breaking that became such a centerpiece of the standard Glashow–Weinberg–Salam model of particle physics.

It was the Higgs boson—named after Peter Higgs, who had predicted the existence of a particle within the symmetry-breaking mechanism—that had caused all the excitement at the CERN press conference on Tuesday, December 13. The experimentalists Fabiola Gianotti and Guido Tonelli, representing respectively the ATLAS and CMS detectors at the LHC, had presented their preliminary results for the new data searching for the Higgs particle. Since a summer conference in Grenoble, about three or four times more data had been collected by the two detectors.

Remarkably, the CMS and ATLAS groups claimed that the Higgs boson mass had been excluded in the energy range from 127–129 GeV to 600 GeV when they combined all the decay channels. This left only a small window of about 12 GeV of energy where the Higgs particle could be hiding. That is the difference between the upper bound limit on the Higgs boson of 114.4 GeV obtained by the LEP2 experiments and about 127 GeV with the new LHC bound for the Higgs mass. At the press conference the week before, after the presentations of the new data, the director-general of CERN, Rolf-Dieter Heuer, claimed that tantalizing new evidence had been obtained from both detectors showing possible hints of a Higgs particle at 124 GeV in the CMS detector, and at about 126 GeV in ATLAS. However, he cautioned that these "bumps" could be statistical fluctuations and might disappear with increasing data and luminosity, or intensity, of the proton–proton collisions.

WHEN IS A BUMP A REAL BUMP?

During the past 20 years, bumps have come and gone in the search for the Higgs boson. Supposed signals of the Higgs would create a lot of excitement and then die down when additional data showed the signals to be merely statistical fluctuations in the data. A standard statistical measure of the significance of observed bumps is the number of standard deviations, or sigmas that can be attached to the resonance bump as it rises above or sinks below the background data. The proton–proton collision observations are represented as curves on a graph, compared with the computer-simulated background. The curve representing the observational data is never smooth because fluctuations in the data occur normally and randomly. These fluctuations are caused by the statistical analysis itself, which never produces a purely nonrandom signal. A 2-sigma result corresponds to about one chance in 20 that the bump in the data is not a statistical fluctuation, but represents something real. A 3-sigma bump is a chance of about one in 20,000 that it is not a fluctuation, whereas a 5-sigma bump signifies that there is only one chance in about 1.7 million that the signal is a statistical fluctuation rather than a real particle resonance. Experimentalists demand that any announcement of a discovery of a new particle must reach the 5-sigma level, removing the possibility that it is caused purely by a statistical fluctuation. Most 3-sigma "discoveries" eventually disappear with the collection of more accurate data. As recently as July 2011, Kyle Kranmer, an experimentalist from ATLAS, reported an excess of events in their Higgs search that reached 2.8 sigma. This was a bump in the data curve at about 140 to 143 GeV. But, then, with the subsequent data presented at the December 13, 2011 press conference this bump was shown to be a statistical fluctuation and had disappeared.

When data are analyzed, it is necessary to put observed events into statistical "bins," which are like storage compartments for data. Bins are a way of organizing the data so that many trillions of events can be analyzed with reduced memory storage on computers. The bins, containing the many data events, are displayed as histogram graphs, with vertical rectangular blocks displayed over a certain range of energy. Experimentalists choose the energy width of each bin to be, for example, 1 GeV, so that any event that falls within this bin is counted. Typically, at a boson energy level of 126 GeV, background events from non-Higgs decay processes will contribute about 400 events per bin. Every statistical background fluctuation of 1 sigma or more could be misidentified as a real Higgs boson signal of 20 events. A 2-sigma fluctuation might be miscounted as 40 extra events. A real Higgs signal could possibly account for 40 events in total, which when boosted by an upward statistical fluctuation, could look like 50 or 60 events. Therefore, a favorable Higgs signal on top of a background fluctuation could be read as a 2-sigma or 3-sigma excess of events. What I have just described accounts for only one bin, whereas several are needed to confirm the existence of a new particle.

The background consists of debris from collisions of other particles. The debris is called *hadronic* because most of it consists of hadrons. Two of the Higgs boson's decay channels are called *golden channels* because they have less hadronic background noise to confuse the picture than the other decay channels. The first is the decay of the standard-model Higgs boson into two photons. The second golden channel is its decay into two pairs of leptons, mediated by the decay of two Z bosons (H^0, a neutral Higgs boson, into a Z boson and a Z^*, which is a virtual Z boson off the "mass shell"[1]; the Z and Z^*, in turn, decay into four leptons).

When considering the two golden channels, we have to be cautious about claiming that a 3-sigma excess of events is a discovery, because there are many bins in the data analysis. The significant question arises: Does a 3-sigma uptick in one of the bins signal a Higgs boson? Or is it just a statistical fluctuation by chance? The possibility that any one bin would fluctuate upward by as much as 3 sigma is less than one percent, but we don't just consider one bin. The probability that one among 25 bins would fluctuate by this amount, or that three

---

1. An important relation in particle physics is the *mass shell*. It is the relationship between the momentum and energy of a particle of a given mass. If we make a plot, with the single component of momentum (mass multiplied by velocity) along the x-axis, and the energy of the particle represented by the y-axis, we must allow for positive and negative values of the movement of the particle to the right and left. The curve showing the energy dependence with momentum is a parabola that has its minimum at the origin of the x- and y-axes. Because the associated energy in the velocity of the particle is quadratic, then for a given value of momentum, the energy is four times greater if we double the momentum. For the two components of momentum, the parabola becomes a shell shape if we rotate it in three-dimensional space. A particle can be called a *virtual particle* if its energy is not inside this mass shell. This virtual particle does not violate conservation of energy because of the Heisenberg uncertainty principle—namely, the violation of conservation of energy happens in such a short time that it cannot be measured.

bins would fluctuate up by something less than three standard deviations each, is far more than one percent. In particular, the probability that you would get a two-standard deviation upward fluctuation just by chance when accounting for 25 bins now reaches a 50 percent probability.

This phenomenon of statistical fluctuation in data analysis is called the *look-elsewhere effect* (LEE). It means that you cannot draw conclusions from just one or two bins of events; you must "look elsewhere" in the range of the probable mass of the Higgs boson to determine whether other bins at other energy levels are ticking up; if they are, this lessens the chance that your original bump is significant. The problem is that the standard model does not predict the mass of the Higgs boson, so its mass–energy can exist in a large range of possible values.

There are two kinds of probabilities. One is called the *local probability* (local p value) and the other is called the *global probability* (global p value). The local probability only takes into account the bins of events at a particular energy value, whereas the global probability takes into account all the possible energy values of where the Higgs boson could be located in the data, and this second probability is where the LEE applies. This kind of statistical analysis plays an important role in many scientific studies. For example, it was used to understand whether the earthquake and hurricane that occurred more or less simultaneously in summer 2011 in the northeastern United States was a statistical coincidence or was caused by a natural convergence.

There are some classical examples that illustrate the LEE. The probability that you will win the lottery is very different from the probability that someone will win the lottery. Similarly, if you are in a room with 100 people, the probability that you will have the same birthday as someone else in the room is small, but the probability that a pair of people will have the same birthday is large. For a given accelerator experiment, the probability that any particular bin in the data will have a 3-sigma excess by chance is very small, but the probability that one of the 25 bins will have a 3-sigma excess is much larger.

If enough data are accumulated by the CMS and ATLAS collaboration that can exclude a Higgs boson outside only a small window of energy, then the significance of the LEE is diminished, and we may be looking at a real discovery.

MEANWHILE, BACK AT THE LAGO MAR AUDITORIUM . . .

In the slide presenting the contents of my talk, I paraphrased a quote from T.S. Eliot's poem "The Love Song of J. Alfred Prufrock": "In the room, 3-sigma bumps come and go/talking of Michelangelo." I was suggesting that during

the past 20 years of experiments searching for new particles, bumps (or resonances) at the 2- and 3-sigma levels appeared in the data, and then disappeared
with increased data and accuracy of the experiments.

I began my talk by discussing the current status of the data on the Higgs
boson search from a theorist's point of view. The detectors at the accelerators
cannot observe a Higgs directly because of its very short lifetime, but can
discover evidence of it by searching for its decay products in various particles of the standard model, such as two W particles, two Z particles, two tau
leptons, two muons, two bottom (b) quarks or two photons. All these decay
channels are predicted for the Higgs by the standard model, if the Higgs
particle mass is known; but, to complicate the issue, other known particles
can also decay into these same pairs. As we have just discussed, the decays
into two Z particles (and then four leptons) or into two photons are dubbed
the *golden channels* because the background of hadronic particles is almost
absent, which allows for a clearer signal for the detection of a Higgs particle.
On the other hand, the backgrounds that we have to worry about for these
golden channels are the photon and lepton backgrounds. However, for the
case of the diphoton decay, the CMS and ATLAS electromagnetic calorimeter
detectors are cleverly designed to select Higgs decays from decays of other
particles into photons and leptons, like electrons and neutrinos. However,
a *dominant* decay channel of the Higgs boson is its decay into bottom and
antibottom quarks. In contrast to the golden channels, this main decay into
two b quarks has to be detected against a background of hadronic decays—
such as the Z boson decaying into bottom and antibottom quarks—which
is millions of times bigger than the Higgs signal to be detected. However,
although the decay of the Higgs into two photons does not suffer from this
large hadronic background, a great many particles can decay into two photons, so this also represents a difficult background that can obscure the decay
of a real Higgs boson.

An important aspect of the decay of the Higgs boson is the "branching
ratio"—that is, the relative percentage significance of each decay channel. The
branching ratio is the decay rate of the channel one is focusing on divided by
the sum of all decay rates of all the possible decay channels. The decay rate
is the inverse of the width of the resonance bump in appropriate units. The
branching ratio for the decay of the Higgs into two photons is only about
0.26 percent compared with the branching ratio of the decay into two b quarks,
which is closer to 60 percent. The signal strength of the decay of the Higgs into
two photons is obtained by multiplying this branching ratio by the number
of Higgs bosons produced, as measured by the production cross-section at a
given energy. Because of the tiny branching ratio of the diphoton decay, this

signal is very small and is hidden by the photon background. However, even though the percentage significance for the two-photon decay is far less than for the bb decay (decay of the Higgs into a bottom and antibottom quark), the lack of hadronic background in the two-photon decay still makes it a much easier decay channel to explore.

I pointed out in my talk that the bumps seen in the possible decay of the Higgs into two photons at about 126 GeV were actually significantly greater in the CMS data than the expected tiny bump calculated by the theorists. In other words, the excess of events representing the decay of the Higgs into two photons was greater than what was expected from theory, which actually suggests that the bump may be a statistical fluctuation. I concluded this part of my talk by saying that we needed more data and significantly greater luminosity to make any decisive statements about whether the bumps were indeed signals of a real Higgs boson.

Another issue I raised is that, in the CMS results, two bumps had been detected, one at 124 GeV and one at about 137 GeV. I had acquired this information from papers from ATLAS and CMS posted on the electronic archive for the December 13 announcement (compare Figures 6.1 and 6.2). The bump

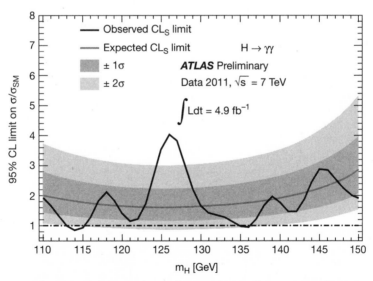

Figure 6.1 ATLAS result for two-photon decay of the new boson, December 2011. The vertical axis shows the ratio of the observed cross-section divided by the cross-section prediction from the standard model; the horizontal axis is the new boson mass in billions of electron volts (GeV). The first, darker band above the line represents 1 sigma, whereas the lighter, top band represents 2 sigma. The plot shows one bump well above 2 sigma, at 126 GeV. © CERN for the benefit of the ATLAS Collaboration

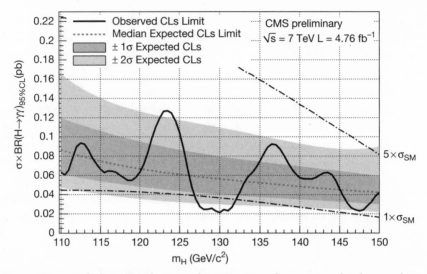

Figure 6.2 CMS diphoton results December 2011. Here there are two data bumps above 1 sigma. The axes represent the same quantities as in Figure 6.1 for the ATLAS results. © CERN for the benefit of the CMS Collaboration

at 137 GeV was almost as big as the one at 124 GeV. So I asked: Which is the real Higgs boson and which is a fluctuation in the data? Will the real Mr. Higgs please stand up?

I ended this part of my talk by stating that the CERN results were inconclusive; no one should claim, on the basis of these data, that the Higgs boson had been detected. However, there were strong hints that a new boson had been discovered and further data obtained at the LHC would confirm whether or not it was the Higgs boson.

At this statement, David Cline, a well-known experimentalist who had been part of the Rubbia team in 1983 that had found the W and the Z bosons, made a loud comment from the back of the hall, saying that I should take into account that the positions of the bumps were the same in the CMS and ATLAS data. I responded that they were not, in fact, in the same position. They could be as much as 2 to 3 GeV apart. Because the width of the bump, or "resonance," that was claimed to be a signal of the decay of the Higgs into two photons, was less than about 1 GeV, one had to be careful about claiming that the bumps were at the same position in energy. The experimentalists were able to resolve the energy position of a resonance to within a few percent. The fact that these two bumps at the different detectors were as close as they were could just be a coincidence. Indeed, the positions of the bumps in the two experiments had better be at the same energy and within the experimental error—which they were not—for the experimentalists to claim they had seen the signal of a real particle.

In the next part of my talk, I listed possible alternatives to the Weinberg–Salam model[2] that did not contain an elementary Higgs particle. These included the possibility that the spin-0 particle was a bound state of Technicolor particles—that is, the supposed Higgs could be a composite of other particles. Another possibility is that new spin-1 particles like the W and the Z existed above 1 to 2 TeV, which could help to remove the violation of unitarity without the Higgs boson. We recall that the violation of unitarity or probability was resolved in the standard model by using the Higgs boson. Without the Higgs, the standard model would have to address the violation of unitarity in a different way.

There were also possible alternative models based on extra dimensions, such as those that occur in string theory. Yet another possibility is a model based on the existence of a fourth generation of quarks, heavier than the observed quarks. This proposal has been promoted for some years by my colleague at the University of Toronto, Bob Holdom. In this model, too, it is possible to avoid postulating a Higgs boson.

Last, I outlined my models of electroweak theory without a Higgs boson, all of which involved a serious overhaul of quantum field theory. Indeed, my electroweak theory based on nonlocal interactions, which I had published earlier in 2011, was a possible candidate.[3] My other model was based on local quantum field theory. It addressed directly the question: Can electroweak theory be renormalizable without a Higgs boson? To restore the necessary gauge invariance of the model required to guarantee renormalizability, I had invoked the Stueckelberg formalism. This was the model I had developed recently, containing only the observed 12 quarks and leptons, the W and Z bosons, and the massless photon and gluon (this model is discussed in Chapter 8).

During the discussion period after my talk, there were several interesting questions. Françoise Englert asked about the size of the troublesome background of the decay of the Higgs boson into two W bosons. I agreed that the background was significant, and caused one to question whether the LHC had yet observed a real Higgs boson event for this channel. I thought that Englert was referring correctly to the fact that the strength of the decay of the Higgs

2. The Weinberg–Salam model was the first to introduce the Higgs mechanism and Higgs boson in a construction of an electroweak model. Glashow had not included them in his 1961 paper. Consequently, physicists have come to refer to the model as the "Weinberg–Salam" model.

3. J.W. Moffat, "Ultraviolet Complete Electroweak Model without a Higgs Particle," *European Physics Journal Plus*, 126, 53 (2011).

boson into two other lower mass bosons was proportional to the mass of these bosons. Therefore, for the Higgs model, the strength of decay would be stronger for heavy bosons such as the W and the Z, and would constitute a significant confirmation of the standard-model Higgs. Perhaps in the future, more events would be observed in this decay channel.

Next, an American veteran of experimental particle physics, Alan Krisch, put up his hand and questioned my skepticism about the LHC having discovered the Higgs. When did I anticipate, he wanted to know, that we would know, truly, that it had been discovered? My answer was: "We have waited 40 years to get to this point of being close to discovering or excluding a fundamental Higgs boson. Surely we can wait a few more months, or even a year, to be certain that the Higgs boson exists or not." He nodded his head in agreement.

COFFEE BREAKS

One of the important reasons that one attends international physics conferences is to meet colleagues who are actively engaged in the research topics with which you are involved. During the coffee break after my talk, I spoke with David Cline. I have always valued David's opinions on experimental results, for I feel that he has always shown integrity in his scientific judgments. David has been involved in large group searches for dark-matter particles for some years. He is currently part of the underground experiment, Xenon100, at the Gran Sasso Pass in Italy, searching for WIMPs. During the coffee break, David expanded his views on the new LHC results, claiming that they were real; it seemed that they had discovered the Higgs boson. I went over my reasons for why I did not yet believe this, and emphasized that we should wait for more data to confirm the result.

I then joined my colleague Philip Mannheim at a table in a courtyard just outside the hotel. The morning was sunny and I enjoyed the warm Florida air after the freezing cold of Canada. We discussed my talk briefly, and Philip concurred that one should show skepticism toward the claim that the LHC had finally discovered the Higgs boson within the narrow remaining window of 12 GeV.

A woman sitting opposite us with her laptop appeared interested in our conversation. I introduced Philip and myself to her. Her name was Monika Wielers. She said that she was one of the analysts on the ATLAS experiment and that there were about 500 of them analyzing the data. I was intrigued by this news and began asking her questions. She said that, as one of the analysts investigating the data, she had not been happy about presenting these results at CERN the previous Tuesday, because the results were very preliminary. She felt that they had been forced into the announcement for political reasons. The director-general of CERN had met with the CERN Council to discuss how they should handle the issue of the Higgs boson politically. In particular, if it turned out eventually that the Higgs did not exist, how would they break this news to the public? The negative news of not discovering the Higgs boson could seriously damage the future of the LHC experiments, for much of the motivation for spending $9 billion to build the machine was to discover the Higgs. If there was a final null result, the media could question the wisdom of spending so much money and time searching for a particle that did not exist. There was a great deal of anticipation in the media, as well as in the physics community, that the LHC would discover the Higgs boson.

I mentioned, as I had in my talk, that the CMS results had shown two bumps, not just the one close to the bump found at the ATLAS detector. Monika smiled and said that the CMS group had more analysts working on the data than the ATLAS group, of which she was part, and that they had devoted more time to their analysis, which could explain why there was more structure in their data. I also questioned her about the fact that, in the ATLAS data, the excess of events, or signal strength, for the Higgs boson decaying into two photons was tiny, and would be hidden inside the background data, whereas the two-bin excess of events in the CMS for the same channel was significantly larger than what the standard model predicted. She said yes, this was a troublesome issue, possibly suggesting that the excess of events was a statistical artifact or fluctuation. The three of us stared at the plot on her laptop, and Philip nodded in agreement; there was something unsatisfactory about all this.

At a later coffee break, I got hold of David Cline again and told him what the ATLAS analyst had said, particularly about the size of the predicted bump for the decay of the Higgs into two photons compared with the size of the observed CMS excess in the data at about 126 GeV. David said that he had talked to a colleague who suggested that the branching ratio for the Higgs decay into two photons should be multiplied by a factor of 10 or more—that is, that the predicted signal strength in the standard model could be wrong by this amount. I said I considered this highly unlikely. Moreover, I pointed out, as in my talk, that the bumps in the CMS and the ATLAS results did not coincide in energy, but differed by as much as 2 to 3 GeV. Considering that the width of the Higgs bump, if it was real, would be less than 1 GeV, then this difference of at least 2 GeV in the positions of the bumps could be significant, and one should be careful about claiming that they are at the same energy. He admitted that the results could be due to statistical artifacts or fluctuations in the data. Indeed, he said that if the experiment provided just a smooth curve without fluctuations, then one could not believe the results, because fluctuations always occur in statistical analyses of data.

Regarding whether there was financial pressure from CERN to announce a discovery, David said that the younger generation of physicists at CERN was naive and did not understand the politics of the situation. He said that CERN had to guarantee that the funding of the machine would continue, particularly in view of the current financial crisis in Europe. Therefore, it was important to make some kind of announcement at the press conference to ensure that funding would continue for the next few years.

At this interesting point in our discussion, we were joined by Lars Brink and Françoise Englert. After filling them in on the main points, I said that I certainly agreed with David on the need for the CERN machine to continue to be funded. It would be a disaster for physics if they cut the funding as the United States had with the supercollider in Texas during the 1990s. However, we must not lose our scientific integrity. I was not willing to compromise and make claims that could eventually turn out to be invalid. Brink and Englert listened impassively.

Later, at another coffee break, I spoke to the ATLAS analyst Monika Wielers again and mentioned that David Cline had suggested that the theoretical calculation of the size of the bump in the Higgs boson into the two-photon decay channel could be increased by a factor of 10 or more. She huffily dismissed this, saying it simply could not be correct. The error in the theoretical calculation could only be about 15 percent, she said, and one could not entertain such a large error. I said it was anticipated that next year, in 2012, the integrated luminosity for the LHC proton–proton collisions would increase to maybe 10 inverse femtobarns, and this could definitely verify or rule out the Higgs boson.[4] She said she was not convinced by this. She felt that they might have to go to 20 or 30 inverse femtobarns before they could reach the magic 5-sigma statistical significance for a bump, which is necessary to claim a real discovery of a particle. Such a statistical significance would amount to saying that there was only a chance of one part in about two million of the signal being a random fluctuation. The statistical significance of the recent LHC results was between 2 sigma and 3 sigma, and therefore could not meet any reasonable statistical

---

4. Cross-sections are measured in units of "barns" (b); a barn is equal to the cross-section of the uranium nucleus, which is 10^{-28} m^2. "Femto" means 10^{-15}, or a thousandth of a millionth of a millionth. Thus, a "femtobarn" (fb) equals 10^{-43} m^2. The "inverse femtobarn" is how many particle collision events there are per femtobarn. The "luminosity" of the proton collisions, measured over time, is the number of collisions per square meter per second. Experimentalists express the integrated luminosity in inverse femtobarns.

requirement of claiming that it was a real signal of the Higgs boson. It was a hint, nothing more.[5]

THE EXPERIMENTALISTS HAVE THEIR SAY

Later on the Saturday of the conference, it was the experimentalists' turn to present their data. Yasar Onel, one of the experimentalists in the CMS group, reported their results. First he discussed the CMS results that appeared to exclude supersymmetric particles, extra dimensions, and other exotica such as mini black holes. Next, Onel presented the CMS results for the Higgs boson search. He went through the data for each decay channel of the Higgs, and finally singled out the decay of the Higgs into two photons. These data produced two bumps, one at 123.5 GeV and one at 137 GeV, and there were also hints of a possible third bump at 119 GeV. The ATLAS data had only revealed one bump in this energy range. I remembered that Monika had said that the number of experimentalists analyzing the CMS data was larger than the number analyzing the ATLAS data and that the CMS analysts were taking longer to analyze the data, and were analyzing it in greater detail. She had suggested that this was the reason that more structure was seen in the CMS data compared with the ATLAS data.

Onel continued that it was important to note that the energies of the bumps in ATLAS and CMS did not coincide or overlap within 2 sigma and therefore it could not be claimed that both occurred at the same energy, leaving open the possibility that the bumps were artifacts or statistical fluctuations. Another important issue, he said, was that the bump in the CMS data at around 123 to 124 GeV had a statistical significance of 1.9 sigma after the look-elsewhere effect was taken into account. This again suggested that far more accurate data were needed to claim that the CMS bumps were a real signal of a discovered particle.

Monika Wielers then presented the most recent results from the ATLAS detector at the LHC. She described how there have been significant exclusions of superpartners in supersymmetry models. Indeed, large swaths of the

---

5. On July 4, 2012, on the basis of between 5 inverse femtobarns and 6 inverse femtobarns of luminosity, CERN announced the discovery of a new boson decaying into two photons, at about a 4-sigma statistical significance. By combining data from this two-photon decay channel with data from the decay of the new boson into ZZ* and four leptons, they were able to claim a 5-sigma discovery. In November 2012, in Kyoto, on the basis of a luminosity of 12 inverse femtobarns, CERN representatives were able to confirm the statistical significance of the new boson beyond 5 sigma. Again, this was based on the combined data from the two-photon decay channel and the ZZ*–4 lepton decay channel.

parameter space of the simplest supersymmetry model of particle physics, called the *constrained minimal supersymmetric model*, had been excluded up to an energy of almost 1 TeV. Moreover, there were new exclusions of extra dimensions of space predicted by various higher dimensional theories such as string theory up to 2 to 3 TeV. Thus, both detectors were, in effect, knocking the ground out from under hundreds of theoretical physicists' feet, in saying that they had not found supersymmetric particles or the higher dimensions of string theory or mini black holes.

Last, Monika described the results of the search for the Higgs boson at the ATLAS detector. Like Onel, she presented the data for the various possible decay channels of the Higgs boson. The two golden channels—namely, the Higgs decaying into ZZ and into two photons—were of particular significance because the decay of the Higgs into a more dominant channel, WW, had serious background problems and the number of data events was not significant. The most significant data was the bump in the diphoton decay channel, which had a statistical significance, or sigma, of about 2.3 standard deviations occurring at an energy of 126 GeV. She showed the important plot of the expected probability of the decay of the Higgs into two photons versus the horizontal axis depicting the Higgs mass—namely, the mass associated with the two photons (see Figure 6.3).

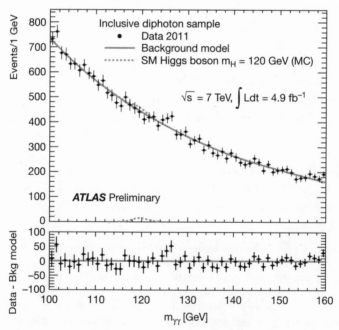

Figure 6.3 Number of events per billions of electron volts for the Higgs decay into two photons. © CERN for the benefit of the ATLAS Collaboration

Two issues of importance appeared from this plot. There was, indeed, a two-bin excess of data events at about 126 GeV that could be the signal of a particle. However, at 120 GeV, there was a little pimple drawn in dashed red depicting the probability of the Higgs boson appearing at that energy level as calculated from the standard Weinberg–Salam model. It was clear from Monika's graphs that, as I had explained in my talk, the observed two-bin events at 126 GeV were much in excess of what could be expected of a standard-model Higgs boson, as shown in the pimple at 120 GeV. I raised my hand and asked her, "Could this not indicate that the excess seen is a statistical artifact, such as a fluctuation in the data?"

Monika walked up close to where I was sitting near the front of the auditorium, smiled, and said, "Yes, I agree with that."

I then suggested that the fact that it was only a slightly more than 2-sigma effect, and that the LEE would reduce its statistical significance to 1.9 sigma, meant that the existence of a Higgs boson could not be confirmed at 126 GeV. I suddenly said in exasperation, "Why is it that this long-sought-after particle, after 40 years, insists on hiding itself in the most difficult place for the accelerators to discover it!"

Monika laughed and said, "I don't know the answer to that question." Returning to address the audience, she then stated that it was too early to draw any conclusions about whether they had seen the Higgs boson in the ATLAS data. Perhaps next year, with a larger luminosity in the proton–proton collisions, and more data, we would be able to decide whether the bump at 126 GeV would increase in significance or just go away.

I pointed out, too, in the discussions after Monika's presentation, that at the July conference in Grenoble, the experimentalists who presented the ATLAS data up to two inverse femtobarns showed a bump with a statistical significance of 2.8 sigma at about 140 GeV. However, this statistical significance was reduced by the LEE. I suggested that the conclusion to all of this next year might be a repeat of last July's ATLAS story. On the other hand, if more data increased the significance of the bump, then we could begin to anticipate that a new particle had been discovered. Time would tell.

Although the Tevatron collider had shut down officially in September 2011, the experimental group was still busy analyzing its data on the search for the Higgs boson and other exotic particles. Two speakers presented the Tevatron data. The first, Todd Adams, concentrated on the results for tests to observe charge-parity (CP) violation—namely, the violation of charge and parity symmetries in bottom quark decays. The standard model did not predict any violation of charge and parity symmetries. The charge quantum number of the quark results from replacing the quark with an antiquark, whereas parity is seeing a particle as an exact mirror image. The CP symmetry consists of making a charge conjugation transformation on a quark (i.e., replacing it with an anti-quark) followed by a parity transformation. Adams left the discussion of the Tevatron Higgs search to the next speaker, Florencia Canelli.

Canelli showed the results for the Higgs search by describing the data from different decay channels, such as the decays into WW, ZZ, and two photons. Nowhere was there a significant excess of events constituting bumps, as had been seen in the CMS and ATLAS data for the two-photon decay channel. I raised my hand and said that, looking at the data, there seemed to be no excess of events above one standard deviation, particularly in the critical region of 115 to 145 GeV. Moreover, this was particularly true of the crucial energy range between 120 GeV and 130 GeV, in which the putative signal of a Higgs boson was claimed by CMS and ATLAS. She nodded in agreement, and said that she didn't believe that the CERN results were conclusive. This raised the ire of the CMS representative, Yasar Onel. There was a sharp exchange between the two, with Onel directing some of his heated comments at me, sitting several rows in front of him.

I then asked the question: "When can we expect to combine all three experiments—namely, the ATLAS, CMS, and Tevatron results?" It seemed to me that this was necessary to reach a final decision on the existence of the Higgs boson, because at this point the "score" was CMS two or three bumps; ATLAS, one; and Tevatron, zero in the critical energy range of 115 to 145 GeV. "Can we expect that the continued analysis of the Tevatron data could produce stronger results?" I asked.

Canelli said that they were indeed continuing the analysis, and she hoped that finally, when all the data were analyzed, they would have a strong enough result to be combined with the data from the two LHC detectors.

HEDGING BETS ON THE HIGGS BOSON

As a member of the audience, I was impressed by the integrity and show of caution on the part of the speakers presenting the LHC and Tevatron data. Even though so much was riding on them finding the Higgs boson—fame, fortune, Nobel Prizes, and, in the case of the LHC, a guarantee of continued funding—the speakers did not exaggerate the possibility of having discovered the long-sought Higgs boson.

In view of the preliminary presentation of the data at the CERN meeting on December 13, I anticipated that there would be a rash of theoretical papers claiming to have predicted a Higgs boson at about 126 GeV. Indeed, among what turned out to be many such papers, one on the electronic archive by physicists at Michigan University claimed that a complicated model based on string theory and supersymmetry predicted a Higgs boson at about 125 GeV.

During the past three decades, there has been quite a history of "observed" new particles; research groups at colliders would report the discovery of a new particle, and subsequent data would eventually show that these new particles did not exist after all. A striking example comes from CERN. Shortly after Carlo Rubbia discovered the W and Z particles in 1983, he announced that he had detected the top quark at 40 GeV. Soon after that, he also announced the discovery of a supersymmetric particle. Further data analysis eventually proved that these two "discoveries" were incorrect, and the evidence for these particles faded away. Much later, the top quark was truly discovered at Fermilab at an energy of 173 GeV. No supersymmetric particles have yet been discovered.

In another instance, in 1984, the experimental group working with the Crystal Ball experiment at SLAC announced the discovery of the Higgs boson at 1.5 GeV. Indeed, ignoring the difference in energy levels, the SLAC plot of the number of events for the decay of the Higgs into two photons looks uncannily similar to the equivalent plots for CMS and ATLAS results presented at the CERN meeting on December 13, 2011. Subsequent additional data and analysis made this 1984 "Higgs discovery" disappear.

A more recent famous example of false discoveries is the finding of a new particle called the *pentaquark* in experiments performed at accelerator laboratories in the mid-2000s. The pentaquark supposedly consists of four quarks and an antiquark bound together—a much different particle than the three quarks making up the proton and neutron. Viewing the plots that were published in the literature between 2003 and 2005, we see a significant excess of events for the pentaquark, reaching almost 5 sigma—the gold-plated standard for the confirmation of new particles. Yet, further data and more detailed analysis again made the new particle disappear. The claimed discovery of the pentaquark resonance was greeted with skepticism, in any case, by the theoretical physicists working on quark models, because the particle did not fit in with the standard, nonrelativistic quark model.

With the advent of more data in 2012, and after the startup of the higher energy LHC in 2015, the bumps in the ATLAS and CMS experiments might similarly disappear. If so, the Higgs boson will be ruled out as an elementary particle up to an energy so high that the standard Weinberg–Salam model would no longer remain viable. In that case, one would expect a surge of theoretical papers claiming, on the basis of one model or another, that the authors had predicted this outcome all along, and we would then enter a long period of theoretical explanations as to why the Higgs boson had not been seen.

Indeed, prior to the December 13, 2011, CERN presentation, papers had already appeared on the electronic archive hedging their bets, claiming that the Higgs boson could be made invisible through various mechanisms, such as the Higgs only decaying into dark-matter particles. It is a truism in physics that theoretical physicists are allowed to speculate and make mistakes in their prognoses of physics. However, experimentalists cannot make mistakes in their claims of discoveries. False claims can seriously damage reputations and careers.

7

Trying to Identify the 125-GeV Bump

On July 4, 2012, CERN announced that the CMS and ATLAS groups had discovered a Higgs-like boson at 125 to 126 GeV. This could turn out to be the standard-model Higgs boson, in which case the standard model of elementary particles will have been completed.

However, if the new boson turns out not to be the standard-model Higgs boson, an alternative to the standard electroweak model must be found. There are already several competing models, and we will enter a phase with the experimental program at the LHC when physicists will be attempting to decide which possible alternative is most viable. One model, my quarkonium resonance model, is based only on the already observed particles in the standard model—namely, the 12 quarks and leptons, the two W bosons, the Z boson, the photon, and the gluon. This model identifies the new boson as a spin-0 resonance composed of a quark and an antiquark.

Another scenario, if the new particle turns out not to be the expected Higgs boson, is that when the energy of the LHC is increased to its maximum, 14 TeV, and the intensity of the proton–proton collisions is increased significantly, then perhaps new particles, including the Higgs, will still be discovered at these higher energies. However, in this case, because of the fundamental problems associated with the Higgs boson, such as the Higgs mass hierarchy problem, new physics beyond the standard model will still have to be found to ameliorate the difficulties with the Higgs boson.

INTERPRETING A BUMP

Because the Higgs boson is so short-lived, we cannot detect it "directly" at the LHC. Its presence can only be inferred by observing the lower-energy particles

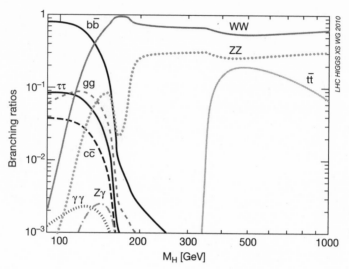

Figure 7.1 Branching ratio magnitudes for possible decay channels of the Higgs boson.
SOURCE: © CERN.

that it can decay into—the so-called *decay product channels*. Detecting a Higgs–like signal within the large backgrounds is like the proverbial search for a needle in a haystack.

There are nine observable decay channels for a Higgs boson with a mass of 125 GeV. The dominant decay channel is into bottom and antibottom quarks (b-bar-b)—that is, about 60 percent of the time, the Higgs decays into bottom and antibottom quarks. The next dominant decay channel is the decay of the Higgs into two W bosons, followed by two gluons. The smallest decay channel is the Higgs decaying into two photons, which only happens about 0.3 percent of the time. In most of these decay channels, there is a very large hadronic, strong-interaction background so that the signal-to-noise ratio is very small, making it difficult to detect the Higgs boson. However, the two-photon and ZZ decay channels are virtually free of this hadronic background, which is why they are called the *golden channels*.

A "branching ratio" for a decay channel is the rate of decay of the boson into its lighter decay products divided by the sum of all the decay channel rates. Figure 7.1 shows graphically the magnitudes of the branching ratios for different decay channels (y-axis) for possible masses of the Higgs boson (x-axis).

From the LHC experimental results up to March 2012, both the CMS and ATLAS detector groups claimed that there were hints of a Higgs particle at 125 GeV in the channel of the Higgs decaying into two photons. At that time, the particle physics community eagerly awaited the results of the LHC startup in April 2012 either to reinforce the hints of a 125-GeV Higgs boson or to exclude it. Then, on July 4, 2012, the 125-GeV bump was statistically vindicated with the CERN announcement of the discovery of a "new boson." The experimentalists are now working on determining whether this new boson is, in fact, a standard-model elementary Higgs boson or some other kind of particle. This entails determining experimentally the spin and the parity of the newly discovered boson—that is, is it a scalar or a pseudoscalar particle? The scalar or pseudoscalar property of the particle is determined by whether the particle changes sign under a parity inversion. That is, when the coordinates x, y, and z are turned into –x, –y, and –z, the scalar does not change sign, whereas the pseudoscalar particle does. Also, into what particles does the new boson actually decay, as opposed to the predicted standard-model branching ratios?

The problem with the Higgs-into-two-photons decay channel is that there is a large photon background (as opposed to hadronic background), and the standard-model predicted resonance at 125 GeV representing the decay of a light Higgs boson into two photons is only about 30 keV. This small signal has to be extracted from the background. Indeed, because of the experimental resolution of the resonance, which currently is about 10^9 electron volts, or 1 GeV, it may never be possible actually to measure this diphoton decay width of the resonance. However, the very sensitive electromagnetic calorimeter detectors at ATLAS and CMS can still measure the total number of events of all particles decaying into two photons, including those from the purported Higgs boson. Experimentalists are then able to extract the Higgs boson decay signal from the total number of events.

We must also measure more dominant decay channels, such as the decay of the Higgs particle into two W bosons, and its dominant decay into bottom and antibottom quarks (b-bar-b). However, with these decay channels, new problems arise because of the large hadronic background. Too many decays of other particles besides the Higgs boson into WW and b-bar-b quarks occur, making it difficult to avoid a misidentification of the Higgs boson. In the case of the b-bar-b decay channel, the hadronic background is more than a million times bigger than the Higgs boson signal would be.

The other golden decay channel is the decay of the Higgs boson into two Z bosons, with masses of about 90 GeV each, which both then decay individually into electron–positron (e+, e–) or mu+ and mu– lepton pairs. The sensitive detectors must identify the final four leptons and measure the masses of the individual pairs of leptons to add up to the masses of the parent Z bosons. If the Higgs mass is below the threshold of two Z bosons at 180 GeV, then one or both of the Zs must be a virtual particle, with a mass less than 90 GeV. In quantum mechanics, a virtual particle is not a "real" particle on the mass shell (refer back to Chapter 6, footnote 1 for a further explanation). As with the diphoton decay channel, the two-Z boson and four-lepton channel is effectively free of hadronic background.

Because the new boson is observed to decay into two photons, it follows from a theorem published by Landau in 1948 and by Yang in 1950 that the new boson has to have spin 0.[1] By using conservation of angular momentum, Landau and Yang showed that an on-shell spin-1 boson cannot decay into two photons, each with spin 1. Therefore, the new boson has to have spin 0 or spin 2. A spin-2 boson is like a graviton. It seems unlikely that a graviton-like particle would be able to fit all the observational data for the new boson, and we must contend with the fact that the new boson is either a scalar or pseudoscalar boson with spin 0. All the experimentally confirmed particles of the standard model have either spin ½ or spin 1, so the question arises: What spin-0 particle can be identified with the new boson? The only elementary particle in the standard model that has spin 0 is the scalar Higgs boson. However, composite bosons made of quarks and antiquarks can have spin 0 and be either scalar or pseudoscalar resonance states.

In two papers, I proposed a model of the 125-GeV resonance bump as a quark–antiquark bound state that I call *zeta* (ζ).[2] This bound state is based on the superheavy "quarkonium" model, which was developed during the 1970s and 1980s on the basis of meson spectroscopy at lower energies, using the nonrelativistic quark model. The term *quarkonium* is borrowed from the word *positronium*, which is a bound state of an electron and a positron. I calculated the partial decay rates of the 125-GeV zeta resonance, which I identified as an electrically neutral pseudoscalar boson, and found that for the zeta resonance decaying into two photons, the result is approximately the same as that expected for a light Higgs boson at 125 GeV decaying into two photons. I found similar results for the channel of the quarkonium zeta resonance decaying into ZZ* and then into four leptons. (Remember, Z* denotes a virtual Z boson.) However, the decays into fermion and antifermion pairs such as b-bar-b quarks and tau+-tau− leptons were suppressed, or much less, compared with the predicted decay of the Higgs boson. This was a significant difference in predictions between the standard-model Higgs boson and my zeta resonance model. In fact, my prediction of the lack of a signal of the new boson decaying into fermion/antifermion pairs such as b-bar-b and tau+-tau− was consistent with the latest data from the LHC.

The message from my quarkonium paper is that it is possible that a bump such as that at 125 GeV in the two-photon decay channel is the result of

1. L.D. Landau, *Doklady Akademii Nauk SSSR*, 60, 207 (1948); C.-N. Yang, *Physical Review*, 77, 242 (1950).

2. J.W. Moffat, "Quarkonium Resonance Identified with the 125 GeV Boson," arXiv.org/1211.2746 [hep-ph]; and "Quarkonium Resonance Model of the 125 GeV Boson," arXiv.org/1302.5583 [hep-ph].

already-established QCD–quark experimental spectroscopy taking place at higher energies. The quark/antiquark pseudoscalar particle is bound by QCD–gluon interactions. If, indeed, the quarkonium interpretation of the new boson turns out to be true experimentally, this would constitute new physics in that it would represent at these higher energies an iteration of well-established quarkonium spectroscopy at lower energies, such as the observed "charmonium" and "bottomonium" resonances at energies of about 3 GeV and 9.5 GeV, respectively. This would certainly call into question the identification of the 125-GeV bump as the elementary standard-model light Higgs particle. At any rate, the paper issues a caution: any definitive statement about a 125-GeV resonance bump being a Higgs boson should be questioned. Only further experimental investigation will be able to decide this issue. Such investigations at the LHC could take one or two years after the machine starts up again in 2015.

By March 2012, there was excitement in the media and on physics blogs about the "hints" of a Higgs particle with a 125-GeV mass. Comments ranged from those who claimed that the Higgs boson had obviously been observed to those who were skeptical about its discovery. It appeared that the majority of experimentalists at CERN and elsewhere at that time were skeptical that they had, indeed, discovered the Higgs boson. The theorists, on the other hand, were either enthusiastic about its discovery or only somewhat skeptical because of the considerable theoretical prejudice in favor of the Higgs that had built up for almost half a century of electroweak physics.

If the Higgs boson was not discovered, then a kind of nightmare scenario would begin to unfold: How can we construct a consistent electroweak theory that permits us to do finite calculations of cross-sections of scattering particles without a spontaneous symmetry-breaking Higgs mechanism? We would have to go back more than 50 years and start all over again with our theories! This would be a wonderful opportunity for younger theoretical physicists who have not spent their whole careers publishing papers on the standard Higgs mechanism model and the Higgs boson. On the other hand, it is a threatening prospect to older physicists who have devoted much of their careers to the standard electroweak theory based on spontaneous symmetry breaking and the Higgs mechanism. However, theoretical physics thrives and progresses on crises. This possible crisis could be compared with the one Max Planck confronted in 1900, when he discovered the formula for black-body radiation, which forced him to conceive of the radical idea of energy being discontinuous, emitted in packages of radiation.

There is an important difference between the hunt for the W and Z bosons in 1983 at CERN and today's search for the Higgs boson. The standard model predicted the masses of the W and the Z; the experimentalists in 1983 had this crucial information when they began searching for the W and Z bosons. Indeed, the experimental discovery of the W and the Z by Rubbia and his collaborators agreed remarkably well with the mass predictions of these particles in the standard model. In contrast, the standard model does not predict the mass of the Higgs boson. Only indirect clues about its mass coming from fits to electroweak data have provided any guidance regarding where to look for the Higgs boson.

Needless to say, as the failure to detect conclusively the Higgs particle at the Tevatron and the LHC continued until March 2012, theorists started to show some nervousness. Papers appeared on the electronic archive speculating about how to explain away the possible absence of the Higgs boson. One popular gambit in theoretical particle physics: if you don't see something that you really believe in, assume that it is "hiding"—make it invisible! How does this work? One can hide the Higgs boson by having it decay into particles that are not easily detected at the Tevatron or the LHC. Or one can provide explanations regarding why the decay of the "hidden" Higgs boson cannot be detected with today's technology.

Amid the flurry of interest in March 2012 over the possible bump at 125 GeV in the two-photon decay channel, there was a problem. The standard model also predicted that the Higgs must decay into leptons, such as electrons and positrons, muons and tau leptons, as well as quarks. These fermion decay channels, which are far more dominant than the two-photon decay channel, had not yet been observed conclusively. So theorists came up with the idea of a "fermiophobic boson," meaning that the Higgs would somehow contrive not to decay into the fermions that one would expect, which again would make the Higgs effectively invisible in these particular decay processes.

Among other possible scenarios for the Higgs decays was the "vectorphobic" model, motivated before 2012 by the absence of a strong signal for the Higgs decaying into the vector bosons W and Z. Such models would appear to be contrived, because they would not agree with the standard-model explanation of symmetry breaking in electroweak theory, nor with the standard-model calculation of the Higgs boson decay into two photons.

It has even been suggested recently that the Higgs boson can decay into dark-matter particles, which of course will probably never be observed! It is difficult, if not impossible, to detect dark-matter particles, as has become clear from the many null underground experiments to detect dark-matter particles such as WIMPs.

The decay of the Higgs into ZZ and WW gauge bosons is important to detect to verify the standard Higgs mechanism, because its coupling to these bosons is necessary to confirm that the Higgs gives mass to the W and Z bosons. Of course, the coupling of the Higgs to the leptons and quarks is also important to observe, so that the Higgs is observed to be responsible for giving fermions their masses. In addition, the Higgs boson cannot decay directly into two photons, because the photons are massless, and a direct coupling of the Higgs to the photons would give them mass. So for the Higgs to decay into two photons, it must do so indirectly through a top/antitop quark loop and a W+, W– boson loop. Therefore, if a strong signal of the Higgs boson decaying into two photons is observed, then this means that the Higgs has to couple to fermions and to the W bosons.

MOVING BEYOND THE STANDARD MODEL

Given the lack of experimental data during the past three decades, theoreticians have been speculating wildly about fundamental issues in particle physics without any experimental restraint on their vivid imaginations. Their goal has been to go "beyond the standard model," inventing new particles and new symmetries—anything, in fact, that would keep the standard model with the Higgs particle intact, but would remove the pesky fine-tuning problems associated with the Higgs boson.

However, since September 2010, experimental restraints have started coming into play from the data pouring out of the LHC, and many of these beyond-the-standard-model (BSM) speculations are suffering a sudden death. If it turns out, in the end, that the bump at 125 GeV is something other than the Higgs boson, then we will have to reinvent the standard model. The question of how particles get mass would have to be reconsidered. This is a problem that requires a deeper understanding of the origins of mass, and how gravity and inertia may also play a role in unraveling this mystery.

There have been many kinds of BSM proposals. Most involve a larger symmetry group than the one on which the standard model is based. The grand unified theories (GUTs) of the 1970s were the precursors of this type of model. The objective of GUTs was to unify the four known forces of nature, including gravity.

These BSM ideas involve increasing the number of undetected particles that are supposedly lurking at high energies not yet reached by accelerators. The most prolific increase of particles is generated by the idea of supersymmetry, which enlarges the symmetry group of spacetime. We recall that with the introduction of supersymmetry, the number of particles is doubled by having a supersymmetric partner for every observed elementary particle. There is also a huge increase in the number of free parameters that describe the properties of these hypothesized superpartners, from about 20 to about 120.

Even if we do solve the problem of how to unify the forces of nature in some grand group structure based on some hypothesized symmetry of particles and fields, this will not necessarily resolve the problem of the infinities encountered in the calculations performed in particle physics. This has been a recurring problem in particle physics during the past six decades and still has to be addressed when attempting to formulate BSM proposals.

One attempt to address the problem of infinities in calculations goes under the rubric of "effective field theory." An effective field theory is not a complete theory that is valid to all energies; rather, it allows one to work up to a certain high energy by means of an energy "cutoff." Take any speculative unified field theory and deal with the problem of the infinities by cutting off the high energies in the calculations, so that you do not have to face what is called the *ultraviolet energy catastrophe*. Physics below about 300 GeV is customarily termed *infrared*, and is in long wave lengths, whereas physics above 1 TeV is in the ultraviolet energy range, with short wave lengths. These short wave lengths correspond to the high energies of particles in accelerators. When the ultraviolet energy extends to infinity in the calculations in quantum field theory, the results of the calculations become infinite and meaningless. This is the ultraviolet energy catastrophe.

Above this ultraviolet energy cutoff, we simply do not understand how to do the mathematics of particle physics, such as calculating the cross-sections of the scattering of particles, and we hope that some future development will rescue us from this impasse. Everything below the energy cutoff is just fine. You can do renormalization theory and all the calculations come out finite, and the particle physicists can happily calculate their cross-sections. However, there is a built-in trap in the weak-interaction theory—namely, without the Higgs boson, using the standard methods of quantum field theory based on local interactions of particles, things can go badly wrong at around 1 to 2 TeV; the scattering of particles is no longer unitary. So even if you put in this energy cutoff at some higher energy—say, 10 TeV—the weak-interaction theory breaks down anyway. In the event that the LHC does not discover any new particles beyond the Higgs boson, and no such new particles exist in nature, then for a light Higgs boson, it can be claimed that the theory is renormalizable and valid to all energies. This means that, in such a renormalizable theory with a light Higgs boson, the cutoff can be made to go to infinity without destroying the calculations of physical quantities such as cross-sections.

The problem with BSM theories in which new particles are claimed to exist is that they may not cure the diseases of weak interactions—namely, the lack of renormalizability of the theories. Exit the Higgs, and you may be faced with this terminal disease. There have been proposals to keep adding in undetected particles, starting at 1 TeV or 2 TeV, and making these particles cancel out the fatal probabilities adding up to more than 100 percent. The problem with this is that more and more particles have to be added into the calculations to keep preventing the unitarity problem from recurring as the energy increases.

However, these arguments rest on the application of perturbation theory to the electroweak theory and to the electroweak theory being valid to all energies. This may not be the case; the theory could become nonperturbative and require a new approach to solving the unitarity problem.

Many of the BSM avenues of research are faced with the problems of lack of unitarity and lack of renormalizability in the weak-interaction theory. If it turns out that there is no Higgs boson, the thinking is that there must be other ways of solving these problems. The culprit in the weak-interaction theory is the charged W massive boson; without the Higgs boson and Higgs mechanism, the W boson destroys the theory because we lose gauge invariance. We recall that this results in the theory not being renormalizable and unitary, unlike QED. Thus, without some radical new way of solving the weak-interaction problem, we are faced with a failed theory. In QED, because the photon is massless and QED has the all-important gauge invariance, the photon's interaction with the electrons avoids all the trouble with obtaining finite calculations of cross-sections and there is no problem with unitarity and probabilities of scattering amplitudes. However, in weak interactions, the fact that the W boson has to be massive to produce the short-range weak interactions spoils this QED picture. Without the Higgs boson, we may lose both renormalizability and unitarity. The Higgs and its accompanying spontaneous symmetry-breaking Higgs mechanism rescued us from this W boson mass problem, and this solution has been entertained for more than 40 years. The whole superstructure of this theory could collapse when you remove the Higgs boson and the spontaneous symmetry-breaking mechanism.

A major part of the standard model of particle physics is quantum chromodynamics QCD, which describes the strong interactions of quarks with gluons using the nonabelian group SU(3). Because the gluons are massless, the theory is fully gauge invariant and also renormalizable. Therefore, the weak interactions, involving the leptons, the W and Z bosons, and the photon, present a greater need for BSM physics than the strong interactions, which involve quarks and gluons. We must somehow succeed in making the electroweak theory finite and meaningful to have reached a successful goal of a standard model of particle physics.

TECHNICOLOR

In solid-state physics, the explanation of how superconductivity works lies in having pairs of electrons (Cooper pairs) binding and condensing together at very low temperatures in metals. In the standard model of particle physics as proposed by Weinberg and Salam, the Higgs boson is considered an elementary scalar particle, not a composite of other particles. But what if the Higgs *is* a composite of other particles, like the Cooper pair condensates in superconductivity? In 1976, Steven Weinberg,[3] followed by Leonard Susskind in 1979,[4] and Savas Dimopoulos and Susskind also in 1979,[5] proposed another way of solving the weak-interaction problem of lack of renormalizability and unitarity. They called their solution "Technicolor." What they proposed was that we replicate the quark-theory QCD at a higher energy with new particles, called *Technicolor particles* because they would carry the quark characteristic called *color*. (Recall that hadrons are observed to be colorless because they are composite particles containing quarks, with three colors that combine, or cancel out, to form a white or colorless hadron.) According to this theory, the Higgs boson is a composite of these Technicolor particles. In the original models, these physicists were concerned only with predicting the masses of the W and Z bosons.

Some versions of Technicolor do not have a Higgs boson at all. Instead of proposing an elementary Higgs boson to explain electroweak phenomena, the Technicolor models "hide" the electroweak symmetry and generate the W and Z boson masses through the dynamics of new, postulated gauge interactions. These new gauge interactions were made invisible at lower energies to fit the experimental data from low energies that do not, so far, reveal these interactions.

The early versions of Technicolor were extended so that one could predict the masses of the quarks and leptons. However, these models ran into trouble in that they predicted neutral current flavor changing in decays of Technicolor particles that violated experimental data. Bob Holdom then proposed a way of avoiding these problems by introducing a type of fine-tuning called *walking Technicolor*.

A notable feature of the Technicolor theory is that the interactions of the Technicolor particles are intrinsically strong interactions, not weak. Although the Technicolor model is a copy of the lower-energy QCD, new strong forces have to be postulated to bind the Technicolor particles together. Therefore, the perturbation methods of quantum field theory used in the standard electroweak model or in QED and in high-energy QCD cannot be used. In low-energy QCD, when it is necessary to explain the confinement of quarks in hadrons,

3. S. Weinberg, "Implications of Dynamical Symmetry Breaking," *Physical Review*, D13, 974–996 (1976).

4. L. Susskind, "Dynamics of Spontaneous Symmetry Breaking in the Weinberg–Salam Theory," *Physical Review*, D20, 2619–2625 (1979).

5. S. Dimopoulos and L. Susskind, "Mass without Scalars," *Nuclear Physics*, B155, 237–252 (1979).

perturbative methods of calculation can fail, and ill-understood nonperturbative methods have to be used, with a large QCD coupling constant.

We do not understand how to perform calculations in which the coupling strength of the interactions is large. The coupling strength of QED, of photons and electrons, is determined by the fine-structure constant, alpha, which is approximately 1 divided by 137, and therefore a small number compared with unity. This small coupling strength of photons and electrons allows us to do the perturbation calculations such that each order of calculation is under control; the magnitude of each order is smaller than the preceding one. For strong interactions of particles in Technicolor models, however, the coupling strength is larger than unity, and we can no longer use perturbation theory. The problems of unitarity and probabilities of cross-section calculations can be resolved potentially in these strong-interaction theories. However, it is difficult to draw any affirmative conclusions about the validity of Technicolor when one cannot calculate anything rigorously with the theory.

Many variations of Technicolor theory have been developed during the past three decades, and the precise data for electroweak experiments at Fermilab have, to some extent, discredited them; these theories do not agree well with the experiments. Somewhat contrived methods have been proposed to counteract the disagreement with the experiments, one of them being Holdom's walking Technicolor model.

Of course, the major problem with Technicolor is that we have to discover all these new Technicolor particles, and the LHC so far appears to have ruled out the obvious lower-energy Technicolor particles such as techni-pions and techni-rhos, which are expected to have masses well below 1 TeV. (Note that, for example, the hypothetical techni-pion, which is the equivalent of the ordinary pi meson, carries color charge whereas the well-known pi meson does not.) The LHC has to explore at high energies whether these Technicolor particles exist. So far, none has been observed, which of course undermines the whole idea of Technicolor. Whenever BSM practitioners are told that a particle they predicted has not been discovered at the LHC, they come up with ways of increasing the mass of the particle beyond the accelerator's current ability to observe it, thus keeping the theory alive. This is happening with Technicolor and supersymmetry today.

One of the chief justifications for the elementary Higgs boson in the standard model is that it generates the masses of the other elementary particles, the quarks and leptons. This happens when the Higgs scalar field has a nonvanishing vacuum expectation value. In quantum mechanics and particle physics, the particles we have observed do not actually exist in a vacuum, except as virtual particle/antiparticle pairs that annihilate one another continually. The vacuum expectation value is the constant nonvanishing potential energy of the particle field in the vacuum. That is, in contrast to other particle fields, the Higgs scalar field energy does not vanish in the vacuum. When Technicolor includes a composite Higgs particle, then it also has a nonvanishing vacuum expectation value.

In the "extended Technicolor model," which predicts the masses of the quarks and leptons, physicists increase the number of gauge bosons in the Technicolor theory that couple to quarks and leptons, thereby proliferating even more the number of particles necessary to make Technicolor a physically viable theory. Again, the LHC will eventually discover or rule out all these hypothesized particles as it reaches higher and higher energies. One of the most severe constraints on all the Technicolor models is the experimental fact that quarks do not change their "flavor" when they transmute from one quark to another through weak interactions. As we recall, the six flavors of quarks make up three generations, with two flavors in each generation. A quark can only decay into another quark in its own generation—that is, without changing flavor. Thus, it is observed that a bottom quark cannot decay into a strange quark. This is an experimental constraint that has to be satisfied in Technicolor models, and it requires a lot of fine-tuning and rather contrived mechanisms to save the model.

ALTERNATIVE COMPOSITE HIGGS MODELS

Besides Technicolor, there are other ways to form composite Higgs models. One is to make the Higgs particle a composite of known quarks and antiquarks, such as the top and antitop quark bound state, called a *top quark Higgs condensate*. This can be compared with the Cooper pairs condensate of electrons in superconductivity. Because the top quark has a mass of 173 GeV, a strong enough force has to be postulated to bind the top quark and the antitop quark into a condensate. A scalar Higgs condensate, made of quarks, replaces the idea of the Higgs boson as an elementary particle.

Several authors have considered the idea that a top quark condensate would get rid of the fine-tuning problems that have plagued the elementary Higgs boson in the standard model. The original idea of a top–antitop condensate was proposed by Yoichiro Nambu and was elaborated further by Vladimir Miransky and Koichi Yamawaki.[6] They used a four-fermion interaction to describe the force that binds the top and antitop quarks to make a condensate. Four-fermion interactions, with quarks and leptons, are not mediated by intermediate gauge bosons such as the W and Z. Such an interaction is not renormalizable, so it would produce unwanted infinities in calculations. The papers on this idea

6. V.A. Miransky and K. Yamawaki, "On Gauge Theories with Additional Four Fermion Interaction," *Modern Physics Letters*, A4, 129–135 (1989).

unfortunately predicted that such a composite Higgs boson would have a mass of about 600 GeV. Clearly, this does not agree with the mass of the top quark, which is now known with accuracy to be about 173 GeV. The condensate would therefore be roughly twice the mass of the top quark, or 346 GeV. The problem with predicting the mass of the Higgs boson from this model arises because the top quark and antitop quark are bound together through a gauge boson, the gluon, which produces a binding energy that has to be accounted for, leading to serious discrepancies with the observed mass of the top quark. We now know, through fitting the precise electroweak data, that a composite top–antitop quark model must produce not a heavy Higgs, but a light Higgs boson, with a mass between 115 GeV and 135 GeV. Moreover, the discovery of the new Higgs–like boson at 125 GeV forces the top–antitop condensate model to incorporate a light scalar Higgs boson, which it has difficulty accomplishing.

Needless to say, particle physicists with their unbridled imaginations have extended the top quark condensate model of the Higgs boson. For example, they included neutrino–antineutrino condensates into the model and other possible quark condensates, thereby potentially lowering the predicted mass of the Higgs particle, making it agree with the global fits of the theory to accurate electroweak data.

A problem with the top–antitop quark condensate is that the top quark decays so rapidly into a bottom quark and a positively charged W boson that the bound state, called *toponium*, cannot exist long enough to be detected. (The lifetime of the top quark is about 5×10^{-24} seconds.) Although toponium has not been detected, it is still considered a state of quarkonium. In contrast, the bottom–antibottom quarkonium state, bottomonium, has been detected at SLAC with an energy of about twice the bottom quark mass—that is, 9 to 10 GeV.

Literally hundreds of papers have been published since 1976 on the topic of creating a physically consistent model of a composite Higgs boson. Recently, physicists have even considered the possibility that the quarks and W and Z bosons are composites of other particles called *preons*, which in turn can produce a composite Higgs boson on their own. The postulate that quarks are made of preons produces new and higher energy scales, which can help to resolve the mass hierarchy problem. These models attempt to remove an unnatural feature of the standard model with an elementary Higgs boson—namely, that the theory has an unstable vacuum. In addition, the standard model with an elementary Higgs boson has the enormous unnatural mass scale difference between the electroweak energy scale of about 200 to 300 GeV and the Planck energy scale of 10^{19} GeV when gravity is speculated to become a strong force. The preon models attempt to resolve this energy hierarchy problem. However, experiments at the LHC during 2011 and

2012 searched for a composite structure of quarks but did not discover any such structure, or any evidence of preons, up to an energy of beyond 1 TeV.

In the event that the LHC does not find any new particles beyond the new boson at 125 GeV, as it increases its energy up to a maximum of 14 TeV, then this mountain of papers suggesting alternative theories will have been produced in vain, which only emphasizes the significance of experimental physics in our quest to understand the nature of matter.

A NON-HIGGS RESONANCE INTERPRETATION

The problem of identifying the new boson at 125 GeV with a spin-0 boson is that, as far as I could ascertain, that would mean that there are only two possible candidates: either it is indeed the elementary scalar Higgs boson or it is some kind of quark–antiquark resonance not formed as a condensate of the top–antitop quarks, which is why I investigated the possibility of a quark–antiquark resonance.

Let us look more closely at this model. I postulated the existence of two new quarkonium resonances, which I named *zeta* (ζ) and *zeta prime* (ζ'). The zeta, which is the lighter of the two, can be identified with the new boson discovered at the LHC, with a mass of 125 GeV. Both the zeta and zeta prime are electrically neutral, and have isospin 0 (isospin singlet), so they have quantum numbers the same as the much lighter pseudoscalar mesons, eta (η) and eta prime (η'), which were first detected during the 1960s.[7]

My new zeta resonances are superheavy quarkonium states, so there is a richer spectroscopy of excited quark/antiquark states associated with their decays, similar to charmonium and bottomonium. The higher excited states of the zeta and zeta prime would be much more difficult to detect than the lowest S-wave state, which has spin 0 and is a pseudoscalar meson—that is, it has negative parity, the same as the eta and eta prime mesons. The zeta and zeta prime interact with one another, and the size of this interaction is measured by a mixing angle of 36 degrees. The mixing angle is calculated using the masses of bottomonium and toponium, which are bound states of bottom–antibottom and top–antitop quarks, respectively. By requiring the zeta mass to be 125 GeV, the predicted mass of the zeta prime resonance is 230 GeV. The strength of the interaction causing the mixing of the zeta and zeta prime is mainly a result of nonperturbative gluon interactions of the top and bottom quarks. This interaction is due to

---

7. Eta and eta prime are isospin singlet pseudoscalar mesons that are made of a mixture of up, down, and strange quarks and their antiquarks. The mass of the eta is 548 MeV, and the mass of the eta prime is 958 MeV.

what is called the $U_A(1)$ axial anomaly first proposed by Gerard 't Hooft during the 1970s to explain the large mass difference between the eta and eta prime pseudoscalar mesons. The zeta meson resonance is a bound state of a quark and an antiquark with an *effective* constituent quark mass of roughly 62 to 63 GeV each, so that the zeta then has a mass of approximately 125 GeV, taking a small binding energy into account.

In this quarkonium model, I stress that the predicted pseudoscalar resonance is not a standard-model Higgs boson nor a quark–antiquark composite Higgs particle condensate. Instead, it is a sequential quark–antiquark resonance at a higher energy than the well-known observed quarkonium resonances at lower energies. In part, I constructed this model to serve as a cautionary message that we should not rush to identify the 125-GeV bump as a standard-model elementary Higgs boson.

8

Electroweak Gauge Theories

Progress in physics is often achieved by viewing popular standard theories from a different point of view. This approach tests the robustness of theories. In my research I find that by constructing an alternative model, I achieve a much deeper understanding of the prevailing standard model. It is necessary to have alternative theories so that the standard model can be compared with them, and also to see which theory best explains the data.

During the early 1990s, I was pursuing alternative theories in cosmology and gravity as well as investigating an alternative to the standard Weinberg–Salam model of electroweak theory, which did not include a Higgs particle.

ALTERNATIVE THEORIES AS FOILS TO STANDARD MODELS

In 1992/1993, I published two papers invoking a *variable speed of light* (*VSL*) to resolve initial value problems in cosmology, several difficulties in our under-standing of events right after the Big Bang.[1, 2] This was an alternative to the stan-dard inflation model proposed by Alan Guth and others during the early 1980s. VSL could explain just as well as inflation the so-called *horizon problem*, the fact that parts of the early universe that were far removed from each other could apparently have the same temperature, without being able to "communicate" with each other. Different parts of the universe could not be in communica-tion through the standard measured speed of light without violating causality, for it would take time for light to cross the universe. Inflation does away with

1. J.W. Moffat, "Superluminary Universe: A Possible Solution to the Initial Value Problem in Cosmology," *International Journal of Modern Physics*, D2, 351–366 (1993).

2. J.W. Moffat, "Quantum Gravity, the Origin of Time and Time's Arrow," *Foundations of Physics*, 23, 411–437 (1993).

this problem by a sudden, exponential growth of spacetime in the very early universe after the Big Bang, whereas VSL increases the speed of light by many orders of magnitude. The VSL model I published predicted a scale-invariant spectrum of matter fluctuations, as did inflation theory, which agreed with observations. Moreover, VSL also explained why the geometry of the universe is observed to be spatially flat. A VSL cosmology required that I modify Einstein's gravity theory and special relativity. In particular, my theory broke the Lorenz invariance symmetry of special relativity. A critical observation that can discriminate between inflation models and VSL is the detection of gravitational waves. Today, VSL is established as a possible alternative to inflation theory; one of the physicists most closely associated with this development is João Magueijo at Imperial College London, who in collaboration with Andreas Albrecht published a paper on VSL.[3]

First in 1992 and then later in 1995, in collaboration with my student Darius Tatarski, I proposed what is now called the *void cosmology*.[4, 5] I pictured that we on earth, and in our galaxy, sit inside a large cosmic void. To describe this—which has also been called the *Hubble bubble*—I used an exact solution of Einstein's inhomogeneous field equations of cosmology to describe a spherically symmetric bubble. Light coming in from galaxies and passing through this void would be influenced by a weaker gravitational field inside the bubble because of the relative lack of matter compared with the outside, which abounded in galaxies and clusters of galaxies. The spacetime inside the void would therefore be expanding faster than the spacetime outside, so that distant galaxies and supernovae would appear to be dimmer than would be expected in a standard homogeneous isotropic universe, such as in the Friedmann–Robertson–Walker standard model. The observer would interpret this dimming of light as evidence for the acceleration of the expansion of the universe. In the void cosmology, the void bubble is embedded in a much larger universe which is not globally accelerating. In contrast, the standard LambdaCDM model says that the expansion of the universe is accelerating and that this is the result of the repulsive nature of "dark energy." In the void cosmology, there is no dark energy, Einstein's cosmological constant is zero or negligible in size, and the expansion of the universe is not accelerating. Numerous papers[6] have been

3. A. Albrecht and J. Magueijo, "A Time Varying Speed of Light as a Solution to Cosmological Puzzles," *Physical Review*, D59, 043516 (1999).

4. J.W. Moffat and D.C. Tatarski, "Redshift and Structure Formation in a Spatially Flat Inhomogeneous Universe," *Physical Review*, D45, 3512–3522 (1992).

5. J.W. Moffat and D.C. Tatarski, "Cosmological Observations in a Local Void," *Astrophysical Journal*, 453, 17–24 (1995).

6. C. Clarkson, "Establishing Homogeneity of the Universe in the Shadow of Dark Energy," *Comptes Rendus de L'Academie des Sciences*, Tome 13, N° 6–7 (2012).

published showing that the void cosmology can fit cosmological data as well as the standard LambdaCDM model, although this still remains a controversial issue today. In a recent paper, astronomers Ryan Keenan, Amy Barger, and Lennox Cowie claim they have observational evidence, from a determination of the average mass density of the local universe, for the existence of a very large void, which contains our Milky Way galaxy.[7]

The third alternative area on which I was working during the early 1990s was modified gravity. One of the biggest mysteries of modern physics and cosmology is dark matter—the need to strengthen the gravitational force with invisible matter in galaxies, clusters of galaxies, and large-scale cosmology to fit the observational data showing much stronger gravity than is expected in Newton's and Einstein's gravity theories. Dark-matter particles have not been detected so far, yet the standard model claims that about 25 percent of matter in the universe is in the form of invisible dark matter. (Almost 70 percent is dark energy; visible matter constitutes only about 5 percent of the total matter–energy budget of the universe in the standard model.) An alternative to dark matter is to modify Newton and Einstein's gravity theories, to strengthen gravity in the presence of ordinary observed matter, which makes dark matter unnecessary. From 1995 onward, I published versions of a modified theory of gravity, which I came to call *MOG* (modified gravity). These were fully relativistic theories of gravity. My collaborators Joel Brownstein, Viktor Toth, and Sohrab Rahvar and I have succeeded in explaining most of the astrophysical and cosmological data currently available in the literature without any dark matter.

A NEW NONLOCAL QUANTUM FIELD THEORY

The problem with mathematical infinities in quantum field theory—our tool for understanding the basic relativistic interactions of matter—has never really gone away during the past seven decades of research in particle physics. Modern refinements of renormalization theory made quantum field theory a more acceptable tool for particle physicists who, unlike Dirac, were prepared to ignore the infinities still lurking in the undergrowth. A nonlocal quantum field theory can raise the issue of whether causality is maintained at the microscopic level of particle physics. Strict causality in quantum field theory demands that, within the tiny distances of particle physics, you will never have an effect before a cause. This requirement is an axiom in quantum field theory

7. R.C. Keenan, A.J. Barger, and L.L. Cowie, "Evidence for a ~300 Megaparsec Scale Under-density in the Local Galaxy Distribution," to be published in the *Astrophysical Journal* (arXiv:1304.28840).

called *microcausality*. The other basic axioms of quantum field theory are unitarity and invariance of the theory under Lorentz transformations (special relativity).

Isaac Newton was not happy with his need to postulate an absolute space and time. Nor was he pleased that the gravity acting between bodies in his theory was instantaneous—the so-called "action at a distance." As Newton himself complained:

> That Gravity should be innate, inherent and essential to Matter, so that one Body may act upon another at a Distance through a Vacuum, without the mediation of anything else, by and through which their Action and Force may be conveyed from one to another, is to me so great an Absurdity, that I believe no Man who has in philosophical Matters a competent Faculty of thinking, can ever fall into it. Gravity must be caused by an Agent acting constantly according to certain Laws, but whether this Agent be material or immaterial, I have left to the Consideration of my Readers.[8]

Experiments have shown that quantum mechanics is a nonlocal theory. This means that in the case, for example, of quantum entanglement between particles such as photons, there has to be an instantaneous communication between the two quantum entangled particles, quite like Newton's gravitational "action at a distance." This violates our understanding of "local" classical physics, which means that distant entities in the universe do not influence us locally. However, in standard quantum mechanics, it is understood that special relativity is not violated, because no actual "information" is communicated between the entangled particles with a speed greater than light. Obviously, there is a contradiction. How can entangled particles communicate, but not communicate information? This is a conundrum that still faces quantum mechanics today.

In Einstein's gravity theory, gravity propagates at a finite speed, the speed of light, and there is no instantaneous action at a distance between gravitating bodies. In a famous paper by Einstein, Boris Podolsky, and Nathan Rosen,[9] they concluded that what Einstein called the "spooky" action at a distance in quantum mechanics meant that it was not a complete theory. In summary, nonlocality means that two objects, such as elementary particles separated in space, can

8. From Isaac Newton, *Papers & Letters on Natural Philosophy* and related documents, edited, with a general introduction, by Bernard Cohen, assisted by Robert E. Schofield (Cambridge, MA: Harvard University Press, 1958), 302.

9. A. Einstein, B. Podolsky, and N. Rosen, "Can Quantum-Mechanical Description of Physical Reality be Considered Complete?" *Physical Review*, 47, 777–780 (1935).

interact without an intermediate agency, and they interact instantaneously. This is Newton's action at a distance. Locality, on the other hand, means that two objects interact through a mediated influence such as a force with a finite speed. In special relativity, this speed of influence cannot exceed the speed of light. Classical physics is strictly local whereas the microscopic objects of quantum mechanics influence one another in a nonlocal manner.

The main raison d'être of *nonlocal* quantum field theory was to create a more natural theory that removed the troublesome infinities that occurred in quantum field theory calculations. The infinities could be traced to the assumption that the particles and fields must satisfy the axiom of *locality* and strict causality at the microscopic level of particle physics. This requirement of locality in quantum field theory would seem to contradict the well-known need for nonrelativistic quantum mechanics to be a nonlocal theory. This situation brings into relief the tension between special relativity, quantum mechanics, and relativistic quantum field theory. In the standard interpretation of the phenomenon of quantum entanglement, spacetime has no influence whatsoever. The whole program of renormalization theory was tied to this idea that particles interact locally at a point in spacetime, which means they cannot affect one another over a region of space.

In Richard Feynman's celebrated paper of 1949 in *Physical Review*,[10] in which he first developed his approach to QED, he traced the origin of the troublesome infinities in the QED calculations to the axiom of microcausality. In technical terms, when he was confronted with the divergent (infinite) quantum radiative corrections in the form of Feynman loop integrals, he introduced into the calculations a "fudge factor" when integrating over the momentum of the virtual particles to infinity. This fudge factor effectively removed microcausality from the theory and made his quantum field theory calculations nonlocal. However, his choice of the "smearing function," which removed the pointlike interactions between the photons and electrons, violated unitarity, gauge invariance, and the Lorentz invariance of special relativity!

In many attempts to develop a nonlocal quantum field theory over several decades, including unpublished attempts by Feynman, the serious obstacles of destroying unitarity, gauge invariance, and special relativity prevented any significant progress. It began to seem that making nonlocal quantum field theory a convincing alternative to the standard local and microcausal quantum field theory was not going to work. The violation of gauge invariance, in particular, generated unphysical negative energy modes in the interaction of particles.

10. R.P. Feynman, "Space-time Approach to Quantum Electrodynamics," *Physical Review*, 76, 769–789 (1949).

In 1989/1990, I began attempting to develop a consistent, nonlocal quantum field theory. Like many before me, including Dirac and Feynman, I was not happy with the infinities in quantum field theory calculations. At the time, I realized that the increasingly popular string theory was, in fact, a nonlocal theory. The string is pictured as a one-dimensional object, whereas a point is a zero-dimensional object, so that when the strings interact, they do not interact at a point. This gave me a clue about how to proceed with quantum field theory in four-dimensional spacetime. Somehow, if string theory was self-consistent, as was claimed, it could not violate gauge invariance, unitarity, or special relativity through the nonlocal interactions of particles. However, the violation of special relativity could only be avoided by situating string theory in more than three spatial dimensions.

In a paper published in 1989[11] based on research done while on sabbatical leave in 1987 in Paris, I hypothesized that there was an infinite "tower" of particles or fields in nature with increasing spin values, from zero all the way to infinity. I demonstrated that when these infinite towers, which I called *superspin fields* or *superspin particles*, interacted with one another, that interaction would be nonlocal. However, this formulation of quantum field theory in four-dimensional spacetime also ran into problems with potential unphysical negative energy modes, resulting from the violation of gauge invariance and unitarity. However, I believed that future work on this theory could possibly produce a physical, self-consistent model of particle physics and unification with gravity by introducing new symmetries for each higher particle spin, which would remove unphysical negative energy modes.

At the time, I chose not to extend the gauge symmetries of this model, and I set aside the idea of infinite spin towers of particles and fields to concentrate on a more standard approach to field theory. I made use of the seminal research on nonlocal field theory published by Russian physicist Gary Efimov during the 1960s.[12] In 1990, I published a paper[13] in which I proposed a method of solving the gauge invariance and unitarity problems in nonlocal quantum field theory by adding compensating field contributions produced by photons and electrons at each term in the perturbation expansion of the fields. These canceled out the problematic gauge-breaking field contributions that had impeded many earlier attempts. I applied this technique to QED and nonabelian gauge theory.

11. J.W. Moffat, "Finite Quantum Field Theory Based on Superspin Fields," *Physical Review*, D39, 3654–3671 (1989).

12. G.V. Efimov, Fiz. Elem. Chastits At. Yadra [*Soviet Journal of Particles and Nuclei*, 5, 89 (1947)].

13. J.W. Moffat, "Finite Nonlocal Gauge Field Theory," *Physical Review*, D41, 1177–1184 (1990).

In this paper, I proposed a way to maintain gauge invariance in a nonlocal electroweak model. I included a scalar Higgs boson with the Higgs mechanism to break the electroweak symmetry. Promoting a nonlocal energy scale (or length scale) as a basic constant of nature, I showed that it was possible to solve the Higgs mass hierarchy problem and make the standard model including a Higgs boson a "natural" theory of particle physics. The paper concluded with a section devoted to a nonlocal quantum gravity that was finite and unitary to all orders of perturbation theory. This paper would prove to be of possible fundamental significance almost a quarter of a century later, when the possible existence of a Higgs-like boson was discovered at the LHC.

In 1990, I invited Richard Woodard from the University of Florida at Gainesville to come to Toronto to give lectures on his work on string theory. Richard was, at the time, an active string theorist, and because there was widespread interest in string theory in the physics community, I wanted to learn more about the subject from him. (Later, Richard abandoned string theory and claimed that the years he spent working on it wasted a significant part of his physics career.) During lunch at the University of Toronto graduate school cafeteria, Richard asked me what research I was currently pursuing. I told him about my superspin field theory ideas and the need for a nonlocal quantum field theory to avoid meaningless infinities in the calculations of particle scatterings. He became upset and told me these ideas were nonsensical and would lead nowhere. Nonetheless, despite Richard's negative reaction, and my worries about it, I published my 1990 paper on nonlocal quantum field theory.

Several weeks after Richard's visit, I received an e-mail message from him claiming that he had had an "epiphany" about my ideas on nonlocal quantum field theory. He understood that the nonlocal field theory offered a way to avoid infinities in QED, and suggested that we pursue these ideas together, in collaboration with my post doc Dan Evens and his graduate student Gary Kleppe. I was certainly willing to go along with this because Richard is one of the world's experts in performing calculations in quantum field theory. Indeed, during the next few weeks, between Toronto and Gainesville, we wrote a long and detailed paper together in which we solved the problems of gauge invariance and unitarity in the new nonlocal quantum field theory while maintaining the Lorentz invariance of special relativity. The paper established that, by introducing the nonlocal behavior of the fields into QED in a special way, we could obtain a theory that was finite to all orders of perturbation theory. That is, the theory did not suffer from the usual infinities in calculations in quantum field theories. In the paper, we extended the theory to nonabelian gauge theories and quantum gravity. It was possible to view the nonlocal extension of quantum field theory as either a regularization scheme simply to make the calculations finite, or as a fundamental theory, depending on whether one kept finite a fundamental energy constant in the nonlocal field theory or let it become infinite at the end of the calculations. We published the paper in 1991.[14]

A special kind of mathematical function called an *entire function* enters into the nonlocal quantum field theory calculations. This function plays a special role in what is called *complex variable theory*, in that it has no singular points anywhere in the finite complex plane. There are only singularities occurring at infinity in the complex plane.[15] The use of entire functions was a key element in my nonlocal quantum field theory, which I had recognized in my 1990 paper and in the paper published in 1991 with Woodard, Evens, and Kleppe.

APPLYING THE NONLOCAL QUANTUM FIELD THEORY TO PARTICLE PHYSICS

After our collaborative effort, I speculated that the nonlocal field theory could be applied to the electroweak unification of weak and electromagnetic interactions, which is normally a local field theory. This formulation would be without a Higgs particle and would generate both a finite electroweak theory and the masses of the elementary particles.

One of the compelling reasons for believing in the standard model of particle physics with its Higgs particle/field was spontaneous symmetry breaking—explaining how the universe went from a supposed state of massless particles to one in which all the elementary particles have distinctive masses, including the zero masses of the photon and gluon. As we recall, according to the widely accepted standard electroweak theory, the Higgs mechanism was responsible for generating the masses of the W and Z bosons and the quarks and leptons. I came up with the idea that the masses of the elementary particles could be generated by natural quantum field theory processes rather than by a Higgs boson. In technical terms, the particle masses could be generated by quantum field theory self-energy calculations without any additional scalar degree of freedom corresponding to the spontaneous symmetry breaking of the vacuum or the resulting prediction of a Higgs particle.

You should understand that, at this time, during the 1990s, it was far from clear that a Higgs boson would ever be detected at the high-energy accelerators,

14. D. Evens, J.W. Moffat, G. Kleppe, and R.P. Woodard, "Nonlocal Regularization of Gauge Theories," *Physical Review*, D43, 499–519 (1991).

15. A complex variable in mathematics extends a real function f(x) to complex functions f(z), where z equals x + iy, where x and y are real variables, and i equals the square root of −1. With the exception of entire complex functions, the function f has singularities in the finite complex plane z. The entire function f must have a singularity at infinity, or else it is a constant.

so there was a significant body of literature investigating how a theory could be developed without a Higgs boson. Moreover, one of the chief motivations for introducing a Higgs boson into the electroweak theory is to guarantee the theory is renormalizable and finite as a local quantum field theory. On the other hand, having a nonlocal quantum field theory that is intrinsically finite removes the need for a local, renormalizable quantum field theory.

In the standard electroweak theory, the mass of the Higgs boson is generated by its so-called self-interaction. In turn, the Higgs boson is responsible for generating the masses of the other standard-model elementary particles—namely, the W and Z bosons and the leptons and quarks. My proposal was that the masses of the elementary particles were all generated by their self-interactions, which meant there was no need for a Higgs boson. In quantum field theory, an elementary particle can interact with itself through fields, which produces a contribution to the mass of the particle. In a nonlocal quantum field theory, this self-energy is finite and it can be made to constitute the total mass of the particle. In a local quantum field theory, you can do this as well, but the self-energy calculation is divergent, or infinite, and has to be made finite by regularization and renormalization methods.[16] Without a Higgs boson in my theory I, of course, had no Higgs mass hierarchy problem, which had plagued the standard model for decades. Another problem that my theory removed by not having a Higgs boson was the extremely large vacuum energy density produced by the Higgs field in the standard model, which seriously disagreed with observation when it was linked to Einstein's cosmological constant.

It was clear to me that because I possessed a finite quantum field theory based on nonlocal interactions of particles in the standard model, I could accommodate a finite renormalization of the masses and charges of the particles. To establish a fundamental nonlocal field theory, it is necessary to introduce a fundamental length or energy scale. Such a constant would take on as fundamental a role as Planck's constant in quantum mechanics. But, removing the Higgs boson from the theory left the unitarity problem of scattering amplitudes still to be resolved.

I published a paper on my nonlocal electroweak theory in *Modern Physics Letters A* in 1991,[17] extending the work published earlier that same year with

16. The idea that the masses of the elementary particles could be produced by their self-energies was originally proposed by Schwinger and Salam independently in 1962, and was developed further by Roman Jackiw, Kenneth Johnson, and Heinz Pagels in 1973; John Cornwall and Richard Norton also in 1973; and Estia Eichten and Frank Feinberg in 1974.

17. J.W. Moffat, "Finite Electroweak Theory without a Higgs Particle," *Modern Physics Letters A*, 6, 1011–1021 (1991).

Richard Woodard on QED. Later, I co-opted my graduate student Michael Clayton into this project, and we did further calculations to try to justify this new electroweak theory.[18] We included a prediction for the mass of the as-yet-undiscovered top quark. We published this second paper in the same year, 1991, in the same journal, *Modern Physics Letters A*.

However, the idea of spontaneous symmetry breaking and the associated prediction of the Higgs boson had, since the early 1970s, become so entrenched in the minds of physicists that an alternative to the Weinberg–Salam model was considered irrelevant, and my published papers on a nonlocal electroweak theory were mostly ignored by the particle physics community. I realized at the time that we were probably 20 to 25 years away from experiments that could decide whether the Higgs boson existed, so it seemed rather futile to continue these speculations and I abandoned my efforts to develop Higgsless models.

BUILDING A NEW PARTICLE PHYSICS THEORY WITHOUT A HIGGS BOSON

Fifteen years later, during the time of the construction of the LHC, the major goal of which was to find or exclude the Higgs boson, I returned to the investigation of whether one could construct an electroweak theory without a Higgs boson. Clearly, the issue of whether the Higgs particle existed would eventually be decided by the LHC experiments. I became curious again about how robust the standard-model electroweak theory was. If the Higgs boson was confirmed definitely to exist, then there would be no need for any alternative electroweak theory. On the other hand, if the Higgs boson were excluded experimentally, then it would be necessary to redesign the Weinberg–Salam model.

The experimentalists at the Tevatron accelerator at Fermilab had devoted serious attention to seeking the elusive Higgs particle before the LHC became operational. So far, the Tevatron experiments had not succeeded in finding a Higgs boson. Now, it seemed to me, was the time to rethink the electroweak theory. Since then, as the data continue to accumulate at an astonishing rate— no longer at the Tevatron, but at the LHC—the possibility that the Higgs boson does exist is becoming real. By July 2012, the hints of its existence were growing stronger, with the announcement of a new boson at 125 GeV. But during the construction of the LHC, things looked quite different.

18. M.A. Clayton and J.W. Moffat, "Prediction of the Top Quark Mass in a Finite Electroweak Theory," *Modern Physics Letters A*, 6, 2697–2703 (1991).

In 2008, Viktor Toth and I investigated further my early ideas from 1991 on how to generate particle masses from a nonlocal quantum field theory.[19] At that time, there were no experimental hints of the existence of the Higgs boson—and if the current evidence for a Higgs boson turns out to be false, there should be renewed interest in the physics community in alternative, non-Higgs models.

Our paper, like the standard model, was based on the premise that you start with a phase in the early universe in which all the elementary particles are massless, including the W and Z bosons. In our model, as in my earlier papers, the particle masses were generated from the quantum field theory self-energy mechanism, in contrast to the standard Higgs model, in which the masses were generated by the interactions of the particles with the Higgs field in the vacuum.

Later, though, I began worrying about the widely held assumption that the universe starts with an unbroken symmetry phase in which the quarks, leptons, and W and Z bosons were massless. Why did this have to be so? A reason for making this assumption in the standard model is that it invokes the basic symmetry of the theory, in this case the group symmetry described by $SU(2) \times U(1)$. The idea is that some mass-generating mechanism breaks this symmetry, which is necessary because the different masses of the elementary particles demand that the symmetry be broken. Yet, I wondered whether the whole idea of generating masses for the elementary particles from either a Higgs–ether field or from complicated quantum field theory calculations was unnecessary. Maybe the particle masses had their origin in some other fundamental physical process, such as gravity or the mysterious origin of the inertial mass of particles?

I decided to rethink the whole electroweak theory from a different angle. I considered the possibility that there never was a phase in the universe in which particles were massless, except for the photon and the gluon. To agree with experimental data, I assumed that the basic nonabelian symmetry group $SU(2)$ was always intrinsically broken in the universe and, apart from possible quantum field theory corrections, the particle masses were simply represented in the calculations by their experimental values. That is, perhaps it was a meaningless question to ask how the leopard got his spots; he simply always had them.

The idea that particles always had their masses may not seem so surprising to a nonphysicist, but since the 1960s, particle physicists have clung persistently to the belief that there exists an explanation for the origin of particle masses. This belief was only reinforced by the publication of the idea that spontaneous symmetry breaking and a Higgs field could explain the origin of masses. Although this explanation could be justified for the W and Z bosons, it was not as convincing for the quarks and leptons because their masses could not be predicted

---

19. J.W. Moffat and V.T. Toth, "The Running of Coupling Constants and Unitarity in a Finite Electroweak Model," arXiv.org/0812.1994.

by their interactions with the Higgs field. Their masses were simply adjusted to fit the experimental data by means of free coupling–constant parameters. One may wonder why particle physicists were so focused on discovering the origin of the masses of elementary particles, while not even considering the origin of, for example, electric charge or the color charge of quarks. Or why were attempts not made to explain the origin of the fine-structure constant, which measures the strength of electromagnetic interactions? In the past, physicists such as Wolfgang Pauli and Arthur Eddington had speculated on the origin of the fine-structure constant, but later generations of physicists showed little concern about the beginnings of anything beyond mass.

I wrote a new paper on electroweak theory that started with the premise that there does not exist a massless phase of fermions and bosons in the early universe, except for the photon and gluon. This paper was published in 2011.[20]

BACK TO LOCAL QUANTUM FIELD THEORY

One of the challenges in doing cutting-edge research in particle physics is to make sure that you have explored all possible avenues when approaching a fundamental problem. I began to have the nagging feeling that I had not sufficiently justified giving up local quantum field theory and its retention of microcausality. So I then posed a new question: Is it possible to develop a renormalizable theory of electroweak interactions using local quantum field theory without a Higgs particle? This required a complete mental turnaround from my nonlocal quantum field theory ideas back to the standard local quantum field theory. Had we missed some key element in our investigations of weak interactions in electroweak theory since the early 1960s? Was there some important feature of local quantum field theory that was missed from the beginning that could avoid the infinities produced by the massive W and Z particles in calculating scattering amplitudes? I appreciated the fact that the odds of reaching this goal were small, considering that many of the best brains in physics had contemplated this problem for more than half a century. Nevertheless, I forged ahead to investigate this question just to satisfy my own curiosity.

The key barrier to finding a successful finite, local electroweak theory was the issue of gauge invariance. I posed the question: Is there some way of reintroducing gauge invariance with massive W and Z particles? Recall that because the

20. J.W. Moffat, "Ultraviolet Complete Electroweak Model without a Higgs Particle," *European Physics Journal Plus*, 126, 53 (2011).

W and Z are massive, they break the required gauge invariance of electroweak theory, rendering the theory unrenormalizable and violating unitarity.

Back in 1938 Swiss physicist Ernst Stuekelberg proposed a way of maintaining the gauge symmetry of a quantum field theory even though the masses of the particles interacting in the theory were put in "by hand."[21] Normally, if you simply put the particle masses into the theory by hand, this breaks the gauge invariance of the theory and it becomes unrenormalizable and violates unitarity. Stuekelberg published his paper in French in a European journal that was not widely known to quantum field theorists at the time. He recognized in his 1938 paper that massive electrodynamics contains a hidden scalar field, and he formulated a version of what would become known as the abelian Higgs mechanism.

His method of retaining gauge invariance even with masses added in by hand works without renormalizability problems for electrically neutral boson interactions such as a hypothetical massive photon or the neutral Z particle of weak interactions; however, when we try to extend this Stuekelberg technique to massive charged vector bosons, like the positively or negatively charged W boson, we again run into difficulties with renormalizability.

In 1962, Lee and Yang published a paper[22] that attempted to unify electromagnetism with the weak interactions. This paper was published a year after Sheldon Glashow's seminal paper introducing the idea that we need a weak neutral current, with its associated electrically neutral Z boson, to complete a unified theory of electromagnetic and weak interactions. Lee and Yang based their quantum field theory arguments on a charged massive vector boson, W, and on the electromagnetic photon fields. We recall that this was some 20 years before the discovery of the W and Z bosons at CERN in 1983. Lee and Yang cleverly used a form of Stuekelberg's gauge invariance without actually referring to his paper, and discovered that under certain constraints the theory could be made renormalizable and not destroy the conservation of probabilities (unitarity). This was two years before the Group of Six discovered the significance of spontaneous symmetry breaking in gauge field theory, and five years before the publication of Steven Weinberg's paper on electroweak unification based on the Higgs mechanism and spontaneous symmetry breaking of the vacuum.

I reviewed the many publications alluding to the problems of renormalizability with the W particle, and despite these problems, in 2011 I attempted to construct an electroweak theory using the Stuekelberg formalism to maintain

21. E.C.G. Stueckelberg, *Helvetica Physica Acta*, 11, 299–312 (1938).

22. T.D. Lee and C.N. Yang, "Theory of Charged Vector Mesons Interacting with the Electromagnetic Field," *Physical Review*, 128, 885–898 (1962).

gauge invariance and renormalizability. I titled the paper "Can Electroweak Theory without a Higgs Particle be Renormalizable?" The idea was to discard spontaneous symmetry breaking, the Higgs mechanism, and the existence of a Higgs particle, and begin with a theory that has intrinsic SU(2) nonabelian symmetry breaking, which means that the universe did not start with a massless phase. The masses of the particles are their experimental masses, which are never zero except for the photon and gluon. The theory was based on the minimal model of the observed 12 quarks and leptons, and the W, Z, photon, and gluon bosons.

The Stuekelberg formalism came with some baggage: extra scalar spin-0 bosons, which have not been detected experimentally as elementary particles. These scalar bosons, like the scalar spin-0 Higgs boson, were an essential part of the Stuekelberg formalism. They were not a problem for interactions involving the neutral Z boson, which is associated with an abelian U(1) symmetry like the photon, because these scalar bosons did not interact with the Z boson or the photon. This theory based on Stuekelberg's gauge invariant formalism was potentially a renormalizable and unitary local field theory.

However, for the W particle in my theory, things were not so easy. The unwanted scalar bosons did couple with the W bosons, causing difficulties with renormalizability and conservation of probabilities (unitarity) in scattering amplitudes. Yet the W "propagator" could be realized within the Stueckelberg formalism, allowing the theory to be renormalizable.[23] The conservation of probabilities was in danger in the Stueckelberg formalism of my electroweak theory because of the unavoidable elementary scalar particles interacting with the charged W boson. But it turned out that the masses of Stueckelberg's scalar bosons could be made large enough that they would be undetected by the LHC even when it is running at its maximum energy of 14 TeV. Above that mass–energy, my electroweak theory breaks down and is no longer renormalizable and unitary. So for energies below this maximum value for the mass of a scalar boson, the theory was fully renormalizable and conserved unitarity. This means that such a theory was only an "effective" electroweak theory, valid up to a high energy somewhere above 14 TeV, at which point the nasty scalar bosons start interacting with the W.

To me this was a disappointment, because I was not able to construct a fully complete, renormalizable electroweak theory based on the standard local quantum field theory, which has been accepted by physicists as the gospel truth for the past seven decades. This was in contrast to my electroweak theory based on

23. A propagator in quantum field theory enables a particle to be created at point A in space-time and absorbed at point B. A particular propagator invented by Feynman (the Feynman propagator) plays a starring role in local quantum field theory calculations.

nonlocal quantum field theory, for which I *was* able to construct an ultraviolet complete quantum field theory valid to infinite energy. At this stage, would I prefer the nonlocal electroweak theory over the effective local electroweak theory? I had to answer this question by investigating carefully the experimental consequences and predictions of both theories.

The problem with an effective theory of the kind I was able to construct using standard local quantum field theory is that it raises the problem of how to "cure" the theory when it breaks down at high energies. The solution would require new physics, presumably in the form of new particles at these higher energies—for example, Stueckelberg's scalar bosons. However, if these energies are above the attainable energy of the LHC (14 TeV), then the whole subject would be cast into limbo until a new accelerator is built that can go to much higher energies.

Let us return to the standard Higgs boson model of electroweak interactions. Is this theory valid to all energies? I believe the answer is a firm no. There are issues of instability of the vacuum with a light Higgs boson, such as the 125-GeV Higgs-like boson discovered at the LHC. Moreover, a scalar field such as the Higgs field suffers from the pathology of what is called the *Landau singularity*, or *Moscow zero*, discovered in QED by Landau during the 1950s. This singularity is intrinsic to a scalar field and will appear at some energy reached eventually by future accelerators. It destroys the consistency of the theory at these high energies.

Perhaps the higher-energy accelerator now being planned at CERN, the ILC (international linear collider), will be able to decide these issues of the breakdown of theories at very high energies. This accelerator will collide positrons and electrons at much higher energies than LEP2, making it easier to discover new particles. This is a different process than at the LHC, which is a ring collider that smashes protons and protons. However, in the event that no new particles are detected at the LHC beyond this current new Higgs-like boson at 125 GeV, then the government funding for higher energy new machines like the ILC may be in jeopardy.

If the new 125-GeV boson turns out to be a light Higgs boson, but the hierarchy problems remain, the particle physics community would still have to be convinced to give up the standard local quantum field theory and consider alternative theories such as my nonlocal quantum field theory. Currently, this seems unlikely. Physicists, on the whole, are a conservative lot and do not give up on their cherished theories easily, which have taken decades to develop. What would be required if the LHC comes up empty-handed beyond the new boson is a "changing of the guard." The older physicists who have devoted their careers to developing standard quantum field theory will not want to sacrifice these ideas, even though they do not provide a complete and successful description of particle physics. To young physicists, on the other hand, this situation would be a rare opportunity to come up with new ideas.

ANOTHER POSSIBLE LOCAL ELECTROWEAK THEORY

I made another attempt to build an electroweak theory based on local quantum field theory. I asked myself the question: Can we generate the masses of the elementary particles from a dynamical field theory symmetry breaking? In contrast to my previous attempt, in which the elementary particles had intrinsic mass from the beginning, this time I began by assuming that there was a massless symmetric phase in the early universe. This phase would contain an exact $SU(2) \times U(1)$ group symmetry, as in the standard model with the Higgs boson.

I studied the 1962 paper about the renormalizability of gauge theories by Abdus Salam, my former professor at Cambridge.[24] This paper came out before the spontaneous symmetry-breaking epiphany by the Group of Six in 1964. In it, Salam cleverly formulated three conditions under which a nonabelian gauge theory could be renormalizable. The first condition required that the "bare" mass—a particle's mass without interactions with other particles and fields—of all the elementary particles in a formalism using perturbation theory was zero. Recall that, in a renormalizable theory based on perturbative expansions of the field quantities, the measured mass of a particle is equal to its bare mass plus its self-energy mass. The second and third conditions, involving some complicated mathematics, could not be met for charged W intermediate vector bosons with a nonzero bare mass in a nonabelian gauge theory. However, they could be met for the massive neutral Z particle, which obeys an abelian gauge invariance.

I formulated my local electroweak theory starting with the premise that the bare masses of the elementary particles, and in particular the masses of the W and Z bosons, are zero. Now I invoked the idea that I had investigated previously—that all the masses of the elementary particles were generated by their self-energies instead of by a spontaneous symmetry breaking with a Higgs boson. Following Salam's paper, I used a method of approximating the electroweak equations based on a technique borrowed from atomic physics and condensed matter physics called the *Hartree–Fock self-consistent procedure*. In the perturbation solution, I chose a new vacuum state that allowed me to calculate the self-energies of the particles through a step-by-step iterative method. I was able to show that, within this scheme, and using the self-energies of the particles and a symmetry breaking of the group $SU(2) \times U(1)$ that incorporated the unification of electromagnetism and the weak interactions, the theory was renormalizable. In addition, a self-energy calculation was able to fit the masses of the W and Z bosons while keeping the photon massless. Moreover, my theory achieved unitarity successfully for the scattering of the W bosons at energies between 1 TeV and 2 TeV.

24. A. Salam, "Renormalizability of Gauge Theories," *Physical Review*, 127, 331–334 (1962).

Now I felt that I had arrived at a satisfactory electroweak theory without a Higgs boson. Within the local quantum field theory formalism on which it was based, it would prove to be self-consistent and agree with accurate electroweak experiments.

However, as experimental results were being accumulated at the LHC, it was becoming clear that evidence was beginning to confirm that a Higgs-like boson had been observed at the CMS and ATLAS detectors. A confirmation of the existence of the standard-model Higgs boson would obviously nullify the need for a model without the Higgs boson. However, the theoretical attempts to find a non-Higgs boson electroweak model produced a "null hypothesis" against which the experimental evidence in search of a Higgs boson could be compared.

It is of historical interest that after Salam had proposed a solution to elec-troweak theory in his paper of 1962, he inserted a footnote explaining that, in a collaboration with Weinberg and Jeffrey Goldstone, they had demonstrated that spontaneous symmetry breaking would produce massless Goldstone bosons, which would prevent a successful completion of electroweak theory. The issue of the massless Goldstone bosons, which cannot exist in nature, had triggered a hiatus in the search for an electroweak theory. Then, in 1964, the Group of Six proposed that gauge bosons such as the W and Z "ate" the massless Goldstone bosons, producing massive gauge bosons. This came to be understood as the Higgs mechanism. It is interesting that Salam got caught up in the spontane-ous symmetry-breaking fever and abandoned his approach of 1962, adopting instead an electroweak theory with a Higgs boson, which, as we recall, he pub-lished in 1968 independently of Weinberg's paper a year earlier.

CONFRONTING THEORIES WITH DATA

Developments in physics often have sociological implications. People devote their entire careers to certain speculative paradigms, only to discover when the experimental data come in that they were wrong from the beginning. Right now, particle physicists are being confronted by data from the LHC, which is, in effect, a killing machine. For example, the whole supersymmetry paradigm appears to be on life support now, with the onslaught of new data coming out of the LHC.

As I often remind my physics colleagues, an elementary scalar particle with spin 0 and positive parity has never been detected experimentally since the construction of the first accelerators during the early 1930s. Possibly the reason is that an electrically neutral spin-0 scalar particle with positive parity (and positive charge conjugation) such as the standard-model Higgs boson has the quantum numbers of the vacuum state. All the other flavor quantum numbers, such as strangeness and charm, are zero in both the vacuum and for the Higgs boson. No other observed *elementary* particle has all these particular quantum numbers of the vacuum. This implies that the Higgs boson is closely related to the properties of the vacuum, such as vacuum fluctuations.

In the observed meson spectroscopy of particle resonances, resonances with spin 0 and positive parity (as well as negative parity) are composed of quark and antiquark states, and thus, in contrast to the standard-model Higgs boson, they are not elementary particles. Elementary scalar particles just seem not to exist in nature. The pi meson is a pseudoscalar particle; that is, it is a scalar spin-0 particle with negative parity. It is not an elementary particle, but a composite made up of a quark and an antiquark. It is amazing that so much of modern physics is based on the idea of an elementary scalar particle (and field) even though such a particle has never been seen, until perhaps now.

Another obvious example of an elementary scalar particle occurs in inflation theory, the most popular theory explaining how to resolve the initial value problems in the early universe in the Big Bang model. This theory has gained enormous popularity since 1981. Yet, it is difficult, if not impossible, to develop a successful inflation cosmology scenario without one of these never-detected scalar particles—in this case, the inflaton. Attempts have been made to use spin-1 particles that do exist experimentally, such as the W and Z bosons and the photon, to develop inflation models, but they have not been successful. One of the most remarkable things about modern physics is the way theoretical physicists often turn a blind eye to experimental facts, such as the experimental exclusion of constrained models of supersymmetry. In a sense, they are busy digging their own graves, and eventually they may find themselves buried with a headstone on which is inscribed: THE ELEMENTARY SCALAR PARTICLE, RIP. On the other hand, if the higher-energy LHC does confirm the discovery of the standard-model Higgs boson, then this would be the first time such a particle has been shown to exist in nature, and it would truly be something to celebrate.

A SCALAR HIGGS BOSON VERSUS A PSEUDOSCALAR MESON

Let us consider some of the prominent features of the standard-model Higgs boson. The Higgs boson couples to other elementary particles proportional to their masses. Therefore, experimentalists have to show that the new boson decays into the most massive spin-1 heavy bosons, which are the W and the Z bosons. Indeed, up until March 2013, the experimentalists at the LHC have apparently shown the decay of the new boson into Ws and Zs, albeit one of the pairs of W and Z bosons is a virtual particle, and the other a real particle, and both subsequently decay into real leptons.

Because the boson that carries the electromagnetic force, the photon, has zero mass, the Higgs boson cannot decay directly into two photons. It must decay through the intermediary of two Ws and a top and antitop quark. The probability for this decay to occur is only about 0.3 percent, yet the CMS and ATLAS groups have detected the two-photon decay, which is one of the golden decay channels.

In addition, the new boson should not be seen to decay into very light fermions such as positrons and electrons. However, it should be observed to decay into much heavier leptons—namely, the tau$^+$–tau$^-$, because the tau lepton has a mass of about 1.8 GeV. So far, the decay into tau$^+$–tau$^-$ has not been substantiated convincingly by the LHC.

The most dominant decay of the Higgs boson into fermions and antifermions is the decay into a bottom quark and an antibottom quark, because the bottom quark has a mass of about 4.5 GeV. The experimental results up through January 2013 do not confirm that the new boson decays into bottom and antibottom quarks. The new boson cannot be detected to decay into the heaviest quark and antiquark pair, the top and antitop quarks, because the top quark has a mass of 173 GeV, and therefore a pair of top quarks would have a mass well above the mass of the new boson at 125 GeV.

My quarkonium model predicts the existence of a resonance called the zeta meson at 125 GeV with spin 0 and negative parity, which must be a pseudoscalar meson. The second, and heavier, resonance in my model is the zeta prime boson, which has a mass of 230 GeV and mixes with the lighter zeta through an angle of 36 degrees. The masses of the zeta and zeta prime bosons are determined by a mixing of the known bottomonium and toponium energy eigenstates. I emphasize that my zeta quarkonium is in no way a Higgs boson or a pseudoscalar Higgs boson. It is quite a different animal.

Let us now compare the predictions of my quarkonium model with the predictions of the Higgs boson model. The ground state decay of the zeta resonance, which is a composite of quarks and antiquarks, is a spin-0 boson with negative parity. That is, it is a pseudoscalar boson. In contrast to the elementary Higgs boson, which decays into two photons through the mediation of a decay into two top/antitop quarks or a W$^+$/W$^-$ loop, my bound state quarkonium, which is not an elementary particle, decays directly into two photons. We recall that a calculation of its decay strength is consistent with the latest observational data and comparable with the prediction of the Higgs boson decay into two photons. Moreover, a calculation of the decay of the zeta boson into a pair of Z bosons, one of which is a virtual Z, yields a result comparable with the Higgs boson prediction. The same is true of the decay channel of the zeta boson into a pair of W bosons, of which one W is a virtual boson. However, the calculated decays of the zeta boson into tau$^+$–tau$^-$ leptons, bottom and antibottom quarks, and charm and anticharm quarks are suppressed, or much less than the decay of the Higgs boson into these particles. Therefore, an important prediction of my model is that the experimentalists should not observe a strong signal of the decay of the new boson into fermion/antifermion pairs, which up until now appears to be the case.

A critical problem faced by the CERN experimentalists is to determine the quantum numbers of the resonance bump they have seen at 125 GeV. Primarily, they have to confirm that the particle has zero spin and positive parity. There are only two possibilities for the spin, for a spin-1 particle cannot decay into two photons and conserve spin, so that the particle has to have either spin 0 or spin 2. Technically speaking, the experimentalists have to determine the angular distribution of the two photons in the decay process: H to two photons or H to ZZ* to four leptons. This is not an easy task, and requires a significant number of particle events to decide the issue.

The data analysis performed up until March 2013 compares the standard Higgs boson scalar model with a pseudoscalar Higgs boson model. This latter model, which I emphasize is not the same particle as my pseudoscalar zeta meson, does not follow directly from the fundamental standard Higgs boson gauge theory. In addition, this effective pseudoscalar Higgs model gives a very suppressed decay into a pair of Z bosons or W bosons, compared with the standard-model Higgs boson. Such an effective pseudoscalar boson model, in contrast to the standard Higgs boson model, is not renormalizable and therefore leads to unwanted infinities in calculations of amplitudes and cross-sections.

The heavy, composite quarkonium model of a 125-GeV pseudoscalar resonance acts as a non-Higgs boson "null" hypothesis for the experimental determination of the spin and parity of the new boson. In particular, the confirmation of the parity of the 125-GeV boson—which determines whether it is a scalar boson or a pseudoscalar boson—is of critical importance in confirming that it is the standard-model Higgs boson.

One of the most difficult problems faced by the CMS and ATLAS collaborations is the analysis of the immense amount of data produced by the proton–proton collisions. The algorithms used to analyze the data have to be able to distinguish between a real signal and the background. From rumors, one gathers that the results of the analysis of the fermion/antifermion decays of the new boson, such as the tau$^+$–tau$^-$ decays, have undergone significant changes since July 2012. One hopes that the analysts are not falling into the psychological trap of simply "seeing" in the data what the Higgs boson predictions require. However, every effort is made by the LHC experimental analysts not to fall into this trap.

As I often say, physics is a brutal business. No mercy is shown to our theories by experimental apparatus such as the LHC high-energy accelerator, which is the way it should be. We are trying to discover how nature works, and nature is indifferent to our ideas and our imagined scenarios of how it behaves. Nature has its own rules, and our goal should be to try to understand nature's laws, which make our universe what it is.

9

The Discovery of a New Boson: Is It the Higgs or Not?

MARCH 2012

The latest developments in the saga of the Higgs search are occurring at the Rencontre de Moriond meeting at the ski resort La Thuile, in the Aosta Valley, Italy, from March 3 to 10, 2012. These annual meetings, organized by the French National Institute of Nuclear and Particle Physics, have been held for 30 years or more. Their purpose is to present the results of high-energy experimental and theoretical investigations from the previous year. The Alpine town of La Thuile is famous for its winter skiing and summer hiking. Subtracting the large number of tourists, and physicists turning up from many different countries to attend the Moriond meetings, the small town has a permanent population of only about 800. The meetings are usually organized so that the physicists can ski early in the day and then attend talks in the late afternoon, which can continue past 7:00 in the evenings. I am not attending this particular meeting, but the slides of the talks are released electronically shortly thereafter.

According to results presented at the Moriond meeting, new analyses of the 2011 data by the CMS and ATLAS collaborations at the LHC have sharpened the standard-model Higgs boson search, but nothing seems dramatically different from the presentation at the CERN press conference in December 2011, when there were "hints" of the discovery of a new boson. On the other hand, at Moriond there are new, interesting results from the Tevatron group at Fermilab. Even though the Tevatron machine shut down in September 2011, the experimental physicists working with the accelerator have been completing the analyses of their 2011 data.

Searching for the Higgs boson consists of tracking down possible decay products of the Higgs particle in the proton–proton collisions in the LHC and the proton–antiproton collisions at the Tevatron. The two so-called golden channels are the decay of the Higgs into two photons and its decay into two Z bosons and then into four leptons. The other, dirtier channels that are difficult to interpret because of background noise are the Higgs decaying into bottom–antibottom (b-bar-b) quarks, charm–anticharm (c-bar-c) quarks, and into two tau leptons (tau–antitau pair). In practice, the probability for detecting the charm–anticharm decay is small because the charm quark is lighter than the bottom quark. The coupling of the Higgs boson to a fermion is proportional to the fermion's mass, so the lighter the mass, the smaller the coupling between the Higgs boson and the fermion, and the more difficult the detection of this decay. Therefore, experimentalists concentrate on detecting the b-bar-b and tau–antitau decay channels.

A significant fact has emerged at this Moriond meeting: The ATLAS and CMS detectors do not see any excess of events in the low-energy mass range (between 115 GeV and 140 GeV) for the Higgs decay in the difficult fermion channels. However, it is important, in the end, to verify that the Higgs boson is seen in these channels because they are the dominant decay channels compared with the golden channels. So far, the LHC results have shown only an excess of events—the so-called hints of the Higgs boson—around 125 GeV in the golden channels.

From the theoretical standard-model calculations, we know that the Higgs boson should decay through all the channels that are being investigated in the LHC and Tevatron detectors. Many people have felt that the Tevatron might have a better chance than the LHC of seeing the Higgs boson in its low-mass range for the difficult and dominant decay channel, the Higgs decaying into the bottom and antibottom quarks.[1] The results being presented at the Moriond meeting by the Tevatron groups show a broad excess of events between 115 GeV and 135 GeV, consistent with a Higgs boson around 125 GeV, as already revealed by the CMS and ATLAS data for the two-photon decay channel. However, this excess being observed at the Tevatron accelerator is only about 2 sigma in statistical significance, which is not strong enough to confirm the existence of a Higgs boson in this mass range. In particular, the Tevatron representative said at the meeting, the data do not show a "peak" in the low-mass range that would correspond to a Higgs resonance. The lack of events in the difficult Higgs decay channels at the CMS and ATLAS detectors does not support the overall evidence for a low-mass Higgs boson. In particular, when the

1. This claim is controversial because the Tevatron has inherently less sensitivity for Higgs searches than the LHC.

ATLAS group showed a combined plot for all channels, the evidence for a Higgs boson dropped from about 3 sigma to 2 sigma. Indeed, the results from the ATLAS group are a bit of a shock. The CMS collaboration already reported a small excess for the two difficult channels (bottom–antibottom and tau–tau)— namely, a one standard deviation—which is indeed small. The reporting by the ATLAS group that they see essentially no excess in either of these channels, and even a deficit in the tau–tau channel, is not good news for the Higgs hunters.

At this point, the ATLAS group results exclude a standard-model Higgs boson in the range of 114.4 GeV (the old LEP bound) to 122 GeV, except for a small window of about 1 GeV wide, centered at 118 GeV. The mass range from 129 GeV all the way up to 600 GeV has now been excluded by the ATLAS and CMS combined data. The only window left for the standard-model Higgs boson to exist is between 122 GeV and 129 GeV, more or less centered around the hinted mass value of 125 to 126 GeV that was last seen in the December 2011 data. The bottom line is that the December hints of a Higgs boson at around 125 to 126 GeV have, from the new data analysis of the difficult fermion–antifermion decay channels, dropped by a full standard deviation or sigma to about 2 sigma.

This means that by now, March 2012, the Higgs search has reached a pause. We have received some positive information and some negative information. However, it appears that we no longer have to worry about a Higgs boson below 122 GeV.

One worrying aspect in the data analysis is that the hints of a Higgs boson at around 125 GeV as reported in December 2011 are only based on 20 percent of the data having been analyzed. A skeptic might be led to believe that these hints of a Higgs boson at 125 GeV in the two golden channels are purely a statistical fluke. The new data analysis presented at this Moriond meeting, which has somewhat weakened the evidence for a low-mass Higgs boson, only strengthens this skepticism.

There has been a great deal of discussion about the Higgs boson in the media and on popular physics blogs. One blog is run by Matt Strassler, a theoretical physicist at Rutgers University. He has expressed a healthy skepticism about whether the standard-model Higgs boson has been confirmed definitely to exist. Many in the media, including blogs, are showing anger and resentment toward any skepticism about the so-called 125-GeV Higgs boson discovery. Indeed, papers are now appearing on the electronic archive (arXiv.org) discussing the Higgs boson at 125 GeV as if the LHC data have already established its existence beyond doubt.

As scientists, we have to strive to make convincing statements on the basis of the data as they appear, rather than on the basis of our emotional beliefs. Of course we hope that the Higgs boson's existence, or nonexistence, will be revealed by new data emerging during the running of the LHC and, indeed, that by the end of the year enough data will have been gathered at a high enough luminosity to establish the existence of the Higgs boson or exclude it. There has even been skepticism about this being the case. Unfortunately, it may take a lot more data to confirm the existence of the Higgs boson through direct searches at the LHC; the background problems must be removed convincingly and the gold-plated 5-sigma excess of events showing a definite peak at the Higgs mass must be achieved.

One interesting result that has been presented at the La Thuile/Moriond meeting is a new measurement of the W mass by the CDF and D0 detectors at the Tevatron machine. This remarkable measurement determines the W mass to an accuracy of 0.02 percent. This, combined with the accuracy of the top quark mass measurement, can limit the mass of a potential Higgs boson, because the formula for the radiative or quantum field theory correction to the Higgs boson mass is sensitive to the W mass. According to a formula derived originally by Martinus Veltman during the early 1970s, there is a logarithmic dependence on the ratio of the Higgs mass to the W mass in the quantum radiative correction. With the new, very accurate measurement of the W mass, the best fit to the Higgs mass is 94 GeV, with an error of +29 GeV or –24 GeV. If this result is taken seriously, a Higgs boson at 125 to 126 GeV would be excluded. However, this result has a statistical significance of only one standard deviation, allowing for a possible two-standard deviation 125-GeV Higgs boson.

JULY 3, 2012

At this time, excitement about a potential discovery of the Higgs boson at the LHC has reached fever pitch. Both the ATLAS and CMS groups will announce the latest results of their 2012 runs at the big high-energy physics conference in Melbourne, Australia, on July 9. Before those talks, CERN will hold a press conference on the morning of July 4, tomorrow, which is Independence Day in the United States. CERN wants any announcement of the discovery of the Higgs boson to take place at its own laboratory.

The speculations about the LHC announcement are rampant; the blogosphere churns with unsubstantiated rumors. The Associated Press published a leaked statement today with the headline "Proof of the Existence of the God Particle." Hours later this title was changed to "Evidence of the Existence of the God Particle." This is perhaps the first time in the history of science that a major news agency corrected the headline of a scientific discovery. The reason is that the experimentalists at CERN are being cautious.

Several days ago, Peter Woit, on his respected blog "Not Even Wrong," quoted a leak from the CMS group saying that they had discovered the Higgs boson. This alerted the ATLAS group to the CMS results before the two groups communicated their final results to one another before the July 4 announcement. This blog leak was not taken kindly by physicists at the LHC, because it allowed potential bias to enter into the analysis of the data. Indeed, according to accounts, the two groups were still analyzing the data up to two or three days before the July 4 announcement.

There is good reason for the experimentalists to be cautious in their announcements. In September 2011, the CERN Oscillation Project with Emulsion-Tracking Apparatus (or OPERA) group announced that they had experimental evidence for neutrinos moving faster than the speed of light. However, in May 2012, that was proved to be incorrect; the dramatic claim turned out to be the result, in part, of a faulty fiber optic cable connection to their GPS receiver. Going back in history to another Independence Day, July 4, 1984, as we recall, Carlo Rubbia, who had discovered the W and Z bosons experimentally the year before, made a special announcement from CERN saying that he had discovered the top quark at about an energy of 40 GeV. This also turned out to be false. Much later, in 1995, the top quark was discovered at Fermilab at about 173 to 175 GeV. Moreover, in 2010, experimentalists at Fermilab announced the discovery of a new particle in the W boson two-jet channel at the CDF that was later invalidated by the D0 detector at Fermilab. This announcement produced predictably another flood of theoretical papers identifying the new particle in some speculative beyond-the-standard model prediction. The CERN group understandably wants to avoid a similar embarrassing situation with the Higgs boson.

An article in *Nature* by journalist Matthew Chalmers published July 2, 2012[2] claimed, on the basis of unidentified sources at CERN, that a new particle had indeed been observed at about 125 GeV, in the decay of this particle into two photons. The claim was that the signal was now close to the 5 sigma, or five standard deviations, necessary for an announcement of the discovery of a particle resonance. But Chalmers cautioned in his article that the CERN experimentalists still had to prove that what they were seeing was a Higgs boson and not an impostor—a new particle that no one had thought of before. Chalmers wrote, "Even as rumors fly in the popular media, physicists have begun quietly cheering at CERN.... 'Without a doubt, we have a discovery,' says one member of the team working on the ATLAS experiment, who wished to remain anonymous. 'It is pure elation!'" As I described in a paper on the electronic archive,[3]

2. M. Chalmers, "Physicists Find New Particle, but Is It the Higgs?" *Nature News* (July 2, 2012).

3. J.W. Moffat, "Has a 125-GeV Pseudoscalar Resonance Been Observed at the LHC?" arXiv. org/1204.4702 [hep-ph].

it is necessary to determine not only the decay products of this putative new particle, but also its spin and parity before declaring it is a Higgs boson. This is not an easy task, and it could take time to resolve. The predictions of the standard-model Higgs boson decay products and the size of the signal observed at the ATLAS and CMS detectors are fairly precise. Rumors are spreading that the observed signal in the two-photon decay channel is about two times bigger than it should be, according to the standard-model predictions.

Despite some cautionary statements about declaring the identity of the resonance bump at 125 GeV, other physics bloggers have already announced that the Higgs boson—or the God particle—has been discovered. However, they have begun to temper their statements as the CERN press conference approaches by saying that there is "possible" evidence for the Higgs boson. In other words, they are hedging their bets. Any bottles of champagne uncorked on the fourth of July at CERN could potentially turn into vinegar.

Peter Higgs is being flown to Geneva to be present at the press conference. The media will be at CERN en masse. North American enthusiasts who want to view what could be a truly historical moment in the history of science must be awake at 3:00 a.m. to view the announcement live on the webcast from CERN. I will not set my alarm for 2:30 a.m., but I plan to watch a repeat later during the day. I predict that the CMS and ATLAS detectors will indeed have evidence of increased strength in the diphoton signal already glimpsed in the 2011 data. However, because of the lack of significant evidence in the other decay channels, such as the WW, tau$^+$–tau$^-$, and b-bar-b channels, I believe they will not announce the actual discovery of the Higgs boson, but will say only that there is strong evidence for its existence.

One of the significant predictions of the Higgs boson is that its coupling strength to particles such as the W and Z bosons and quarks and leptons is proportional to their masses. Therefore, its coupling strength to the W boson must be observed as a significant enhancement of a signal of the Higgs decaying into W bosons at about 125 GeV. So far, there have been fewer events than expected in this decay channel observed at both the Tevatron and the LHC. Not to be left out of the excitement about the discovery of the God particle, the Tevatron group posted a long announcement online yesterday, July 2nd, claiming to see a broad enhancement of about 3 sigma of a Higgs boson decaying into bottom–antibottom quarks at about 125 GeV. Puzzlingly, however, the Tevatron group also sees a broad enhancement in this same decay channel above 135 GeV. Again, this discrepancy in energy levels raises the concern that these enhancements may be the result of statistical fluctuations in the data and problems with calibrating the large backgrounds. Also in its online announcement, the Tevatron group states they see a lack of evidence for a Higgs signal in the WW decay channel. These announcements have been made on the basis of further analysis of the Tevatron data, up to an integrated luminosity of about 10 inverse femtobarns, after the machine was shut down in September 2011.

JULY 4, 2012 (INDEPENDENCE DAY)

I wake up at 2:30 in the morning without any help from my alarm clock. Despite my drowsy state, I decide to watch the live video from Geneva of the scientific seminar with the two talks announcing new results in the search for the Higgs boson. The video streaming from Geneva is of excellent quality. Before the start of the conference, the cameras pan the main auditorium at CERN, which is packed with dignitaries. Evidently younger physicists have been standing outside all night, attempting to get there early in the morning to find a seat. At one point, Peter Higgs enters the auditorium and greets Françoise Englert. They both look ebullient, anticipating the new results. The director-general of CERN, Rolf-Dieter Heuer, makes a brief announcement and then introduces the first speaker, Joe Incandela, who describes the CMS results.

Incandela seems quite excited and nervous. After a preliminary discussion of the workings of the CMS detector, he moves on to the results of the 2012 runs. These data result from a beam energy of 7 TeV and 8 TeV, and an integrated luminosity of about 5 inverse femtobarns, which is close to the beam intensity or luminosity of 4.7 inverse femtobarns from the 2011 experimental runs. He claims that combining the new 2012 data with the 2011 data increases the sensitivity for the Higgs search by about 20 percent. Combining the two years of data effectively doubles the amount of available data. Incandela concentrates on the decay channel of the Higgs decaying to two photons, which, except for the problem of the photon background, provides the cleanest data because of the little or no hadronic background to be concerned about. He discusses the data from other decay channels, in particular concentrating on the decay of the Higgs into two Z bosons, each of which then decays into a lepton and an anti-lepton. The signal to look for in that channel is four leptons, such as electrons and muons, coming from the neutral Z bosons. The signal for this decay channel does not show significant excess of events, but the result is still consistent with a Higgs boson.

For the "golden" diphoton decay channel, Incandela announces they measured the mass of the resonance bump as 125 ± 0.6 GeV. He says that, in this channel, the data reached a statistical significance of just above 4 sigma, whereas for the two Zs decaying into four leptons, the statistical significance was just above 3 sigma, less than the gold standard of 5 sigma. By combining the decay events in these two golden channels, Incandela states dramatically, they have reached the magic 5-sigma statistical significance. This warrants the claim that they have discovered a new boson. At this statement, the audience applauds loudly.

The next speaker is Fabiola Gianotti from the ATLAS group. She goes straight to the issue of how they improved the data analysis techniques, and then comments on the amazing work done by the worldwide computer grid that has been analyzing the data for only a few weeks, with a cutoff of June 15. It is amazing how the many hundreds of analysts managed to crunch through the huge volume of numbers to produce the remarkable result that she was about to announce. She, too, concentrates in her talk on the two golden channels: the Higgs decay into two photons and its decay into two Zs and then four leptons. Working through slide after slide of detailed analyses, showing how they separated background noise from the signal, Gianotti finally announces that with the combined data from 2011 and 2012 for the two golden decay channels, they have a 5-sigma or five standard deviation confirmation of a new particle at a mass of 126.7 GeV. The five standard deviations means there is only about a one in two million chance for this result to be a statistical fluctuation. This is a significant increase from the 2011 runs, which had only reached about 3 sigma. Again, when this slide is shown, the audience erupts into boisterous applause.

Watching the slides appear one by one on my computer, I take note of three facts, emphasized by Gianotti. The first is the signal strength of the new boson's decay into two photons. This is measured by multiplying the total production cross-section of the new boson by the predicted branching ratio of the Higgs decay channel producing two photons, which turns out to be about two times larger than predicted by the standard Higgs boson model.[4] If this discrepancy continues to hold up, then the LHC experimentalists will have to question whether this new particle is a standard Higgs boson, some variant of the Higgs boson, or a new kind of particle that has not yet been observed experimentally. However, Gianotti avoids commenting on this issue because her role at this stage is just to provide the experimental results, not to give theoretical interpretations.

The second issue I note is that the measured mass of the new particle at the CMS is different from the one at ATLAS by 1.5 GeV, which could be a significant difference. This was a problem with the results of the 2011 experimental data as well, and it was hoped that this mass difference would go away with the new results. In my opinion, this difference could just be the result of experimental error or statistical variance, and it will likely disappear with new data.

---

4. This branching ratio is measured by the ratio of the predicted partial width of the diphoton resonance bump divided by the total predicted width of the resonance resulting from all the particle decays.

The third issue I note is that of the other decay channels; the Higgs decay into two tau leptons shows a *deficit* of events at 125 GeV—not a spike in the data, but a dip. This result, too, is not consistent with the standard-model Higgs boson predictions.

As I listen to Incandela and Gianotti make their announcements, I think about the next steps to come. To establish that this new particle is the Higgs boson, experimentalists will now have to determine the quantum numbers of the particle, such as the spin and parity, and attempt to obtain a precise determination of all the branching ratios for the other decay channels, in addition to those for the golden channels. When these experimental results are finalized, only then can we say definitely that the particle is a Higgs boson.

Recall that the standard-model Higgs boson is an elementary scalar particle with spin 0 and positive parity. Until now, such a particle has never been observed experimentally. However, there have been observations of mesons with intrinsic quark spin 1 for the quark–antiquark bound state, with the quark spins aligned parallel to one another, and with an orbital angular momentum of 1, which look like a resonance with spin 0 and positive parity. Such a scalar particle is composed of a quark and an antiquark, whereas the standard-model Higgs boson is an elementary particle with no constituent elements inside of it. The quantum numbers S (spin) and L (angular momentum) for quark–antiquark mesons are important to determine experimentally. What is remarkable is that, if indeed these new results from the LHC prove to be a Higgs boson with the quantum numbers of the vacuum or ground state, then this will be the first such particle ever observed in nature. It will be in a family of one.

The final comments by Heuer after the talks do not go quite so far as to say that they have definitely discovered the Higgs boson. He says, "As a layman, I would now say, 'I think we have it.' Do you agree?" The receptive audience again erupts with laughter and applause.

After almost 50 years of particle physics with a hypothesized Higgs boson, and the efforts over 40 years by many thousands of experimentalists to discover this elusive particle at the accelerators, it is clear that the audience is biased heavily toward believing this resonance bump is indeed the Higgs boson. If so, this would be a big coup for the LHC and CERN, and would seem to justify the $9 billion spent on the experiment so far.

JULY 9, 2012

After the televised scientific seminar, there was a press conference at CERN, with a panel of experimentalists moderated by the director-general, Rolf-Dieter

Heuer. There was a large collection of international journalists from all the prominent newspapers, television stations, science journals, and magazines. Peter Higgs entered the auditorium for the press conference surrounded by photojournalists flashing their cameras. Amusingly, he himself was a caricature of the most popular description of how the Higgs boson gives mass to other particles. As soon as he entered the room, he attracted a crowd of journalists and physicists and, moving slowly along, seemed to impart mass to them. He soon joined Françoise Englert in the audience.

In response to many questions from the journalists about whether the new boson that had been discovered was the Higgs boson, the experimentalists on the panel were reticent about saying this was the case. Eventually, after the cool responses of the experimentalists to this question, Englert raised his hand and asked how long it would take them to make the decision that it was really the scalar particle that they were looking for, that fitted into the standard model. Fabiola Gianotti, who appeared to be an objective and level-headed experimentalist, said that it could take several years before a final decision was reached. Heuer echoed that it could take three to four years to confirm whether it was the Higgs boson. He then said in an off-hand way that it would not be decided by the end of 2012. This meant we would have to wait until 2015 when the LHC would start up again at 13 TeV. However, he also stated that they would continue running the machine 2 to 3 months longer than planned during this run to collect as much data as possible.

Despite the cautions from Heuer and the members of the panel, others present at the press conference had no qualms about expressing their enthusiasm:

Michel Spiro, who is president of the CERN Council, said, "If I may say so, it [the discovery] is another giant leap for mankind."

Françoise Englert declared, "I am extraordinarily impressed by what you have done."

Gerald Guralnik enthused, "It is wonderful to be at a physics event where there is applause like there is at a football game."

Peter Higgs mused, "It is an incredible thing that it has happened in my lifetime."

Getting into the spirit of things, Heuer concluded, "Everybody that was involved in the project can be proud of this day. Enjoy it!"

Almost immediately, and ignoring the more cautious stance of the experimentalists, the blogosphere went hysterical about the "discovery" of the Higgs boson. Even at my own institute, the claim was emblazoned on our home page: "*The Higgs boson, sought for decades, has been discovered. What does that mean and where do we go from here?"* Needless to say, I felt disturbed that the objectivity of the physics had been subverted, because according to the experimentalists at CERN, we could not yet say definitively that the Higgs boson had been discovered. Yet the physics blogs were already speculating about who should get the Nobel Prize and when. Indeed, at "Not Even Wrong," Peter Woit proposed that the prize should be given to the experimentalists first, shared among the CMS and ATLAS groups (a total of about 6,000 physicists!). Indeed, in Woit's blog he discussed the possibility that the experimentalists should receive the prize this coming October, only a few months after the discovery of the as-yet unidentified new boson. However, nominations for the prize would have to have been submitted the previous January, to follow the conventional Nobel procedure.

Also on the blogs, a consensus is now being reached that, among the five or six living theorists who came up with the idea of the Higgs mechanism in 1964, Englert, Higgs, and Kibble should be the three to be awarded the prize the following year. Hagen and Guralnik, as well as Anderson, are being left out, for reasons that are unclear. Indeed, Gerry Guralnik was one of the first to consider spontaneous symmetry breaking to be a fundamental part of relativistic particle physics. He had to weather a lot of tough criticism. Legend has it that when Guralnik gave a talk in Munich about this research, the celebrated Werner Heisenberg stood up in the back of the auditorium and called him an idiot.

Historically, the Nobel Prize for the discovery of the W and Z bosons in January 1983 was awarded the next year to Carlo Rubbia and Simon Van der Meer, both experimentalists at CERN. Moreover, the 1979 Nobel Prize awarded to Glashow, Weinberg, and Salam occurred four years before the actual discovery of the W and Z bosons and, of course, the Higgs boson, an integral part of their electroweak theory, had not yet been discovered.

Always lurking in the background of this story is Rubbia's unfortunate announcement in 1984 that he had "discovered" the top quark at an energy of about 40 GeV. This announcement at CERN via a press conference was held, coincidentally, on July 4, 1984. This "discovery" was based on 12 events for the decay of the W boson into a top and an antibottom quark. There were two experiments at CERN that could detect this decay, at an energy of 630 GeV, in the proton–antiproton collider. The UA1 detector observed 12 events against a background of three and a half events. The eight or nine events above background were enough to produce a resonance peak of just more than 3 sigma.

This was considered a double-header triumph for CERN and Carlo Rubbia, coming so close after the discovery of the W and Z particles. On

July 12, 1984, the editor of *Nature*, John Maddox, published an article with the title, "CERN Comes Out Again on Top," commenting on the discovery of the top quark by the group under the leadership of Rubbia.[5] He began the article with the statement, "The Matthew principle—'to him who hath should be given'—is working in favour of CERN..." The article also stated, "The new development at CERN follows almost exactly along the lines expected." However, when more data were collected, it was discovered that the QCD background calculations were incorrect. They underestimated the background, leading to the announcement of the false claim. Moreover, the competing UA2 detector group in 1984 at CERN could not confirm the observations of the UA1 group. This only proved again that the roads of particle physics are paved with the tombstones of 3-sigma events, fluctuations, and unexplained effects.

Once I had time to digest the contents of the July 4 scientific talks by Gianotti and Incandela, the following became clear: The signal for the Higgs boson decay into two photons had gone from about 3 sigma to 4 sigma in strength. The signal in the other golden decay channel—Higgs into ZZ*, which then decay into four leptons—had reached a strength of about 3 sigma. When they combined these two channels, they got a signal strength of the gold-plated 5 sigma for a Higgs boson at a mass of 125.5 GeV at CMS and 126.5 at ATLAS. An important fact that confirmed that CERN had actually discovered a new boson at about 125 GeV was that both groups, CMS and ATLAS, had detected a 5-sigma signal at about the same mass. Yet, there was no official combination of the ATLAS and CMS results announced at the seminar and press conference on the fourth of July.

Remember that the signals in the other channels, such as Higgs decaying into WW, b-bar-b, and tau–tau, were much weaker. In fact, the tau–tau channel decay had a signal of zero, with no examples of it during the collisions. In addition, the two-photon channel decay signal was about twice the predicted size from the standard model, corresponding to about a 2.5-sigma deviation from the standard-model prediction. So it could be said that the experimental result was consistent with a Higgs boson, within 2.5 to 3 standard deviations. However, as the experimentalists stressed at the press conference, the spin of the resonance and the all-important parity were unknown. Therefore, it could be that an "impostor," as Gianotti put it, is simulating a Higgs boson.

In contrast to the experimentalists, the theoretical particle physics community was heavily biased toward the LHC having discovered the Higgs boson.

5. J. Maddox, "CERN Comes Out Again on Top," *Nature*, 310, 97 (1984).

This is understandable because of the almost 50 years of publications on the Higgs boson by theorists, and the many textbooks used in universities in which the Higgs boson and the idea of spontaneous symmetry breaking play a prominent role. However, this bias on the part of the theorists had not deterred the experimentalists from taking a cautious and objective view of the situation until more data were collected and it could be established beyond a doubt that the new resonance is a Higgs boson.

In view of all this excitement, I couldn't help pondering the technical and fundamental consequences for particle physics of finding a standard-model Higgs boson as an elementary scalar particle. The Higgs mass hierarchy problem, and the absurd fine-tuning prediction of the vacuum density resulting from the Higgs field, are not going away. Indeed, given the lack of evidence in the LHC data for the minimal supersymmetric standard model (MSSM), which could solve the Higgs mass hierarchy problem, as well as the gauge hierarchy problem and its extreme fine-tuning, the discovery of a Higgs boson on its own could be something of a nightmare. Any new particles discovered with a mass of 1 TeV or more, pointing toward a BSM scenario could, because of the scalar nature of the Higgs boson, still make the standard model untenable. Such a new heavy particle mass produces corrections to the 125 GeV observed mass that require a fine-tuning correction of 33 decimal places if the energy cutoff in the calculations is the Planck energy. Moreover, there is another issue with the standard model and the Higgs boson—namely, the discovery that the neutrinos have a mass. The origin of the neutrino masses is a mystery, and if no new heavy neutrinos are discovered experimentally, such as "sterile neutrinos," which do not interact with ordinary matter, then the standard model with the Higgs boson may not be a renormalizable theory, so the original justification for having a Higgs boson is not so strong.

It still seemed to me that the neatest solution to these somewhat catastrophic consequences of the Higgs boson would be avoided most easily simply by not having a Higgs boson! My proposal that the bump could be a quarkonium resonance, which I had been thinking about for some time, is not the only alternative model suggesting that the discovery is not a simple standard-model Higgs boson, but a more complicated beast. For instance, several physicists have suggested that two Higgs particles, or maybe even three, interact to produce the observed data. Another alternative predicts the existence of new charged and neutral particles to boost the size of the two-photon decay channel signal in the theory to agree with experiment. But, if the future data at the LHC continue to support the discovery of an elementary Higgs boson, then these possible solutions to the Higgs boson problems, including mine, will have to be abandoned.

JULY 11, 2012

It is critical to distinguish the spin and parity of the elementary, fundamental Higgs boson from what can be called a *Higgs impostor*. Particle theorists recognized during the 1990s and early 2000s that the spin of the standard Higgs boson had to be zero and its parity had to be positive. That is, it has to be a scalar boson. The scalar Higgs and an impostor can be distinguished experimentally through careful analysis of the angular correlations in their decay products.[6] We know that because the 125-GeV boson has been observed to decay into two photons, and into a Z and a virtual Z boson, that the spin of the newly discovered boson has to be either spin 0 or spin 2. It cannot be spin 1, according to a theorem published years ago by Landau and Yang, who showed that a spin-1 vector particle such as a photon cannot decay into two spin-1 vector particles. Particle physicists prefer the spin-0 boson over spin 2, because a spin-2 particle would have the characteristics of a graviton, and most physicists do not believe that the decay of such a spin-2 particle has been seen in the data.

A distinct signature that distinguishes between a scalar spin-0 boson and a pseudoscalar spin-0 boson is that, at high enough energies, the decay products of a scalar boson become longitudinally polarized predominantly, whereas the decay products of a pseudoscalar boson become mainly transverse in polarization. These longitudinal and transverse polarizations constitute the degrees of freedom of the propagation of a particle in space. The longitudinal polarization or degree of freedom means that the particle has a spin direction oriented along its path of motion, whereas transverse polarization means that the particle waves are perpendicular to the direction of motion of the particle. Experimentalists are able to detect these polarizations in the data.

It was during the week after the CERN press conference that I finished constructing my own Higgs impostor model, the quarkonium resonance called *zeta* that I have referred to several times in this book. This model can be used to discriminate between a Higgs boson and a non-Higgs boson. This can be done by a careful comparison of the decay rates and branching ratios of the standard-model Higgs boson and those predicted by my impostor model. In particular, the impostor boson is a pseudoscalar particle with negative parity

---

6. When the spin-0 boson decays into either two photons or two Z bosons and then into four leptons, the decaying particles can be analyzed in terms of the angles at which they deviate from each other in their decays. For the case of the two pairs of leptons in the ZZ decay, five angles can be constructed from the data. The results of such analyses demand precise, statistically significant data, which have not yet been obtained and may not be available before 2015/2016.

compared with the positive parity of the standard-model scalar Higgs boson. Moreover, the impostor boson has spin 0, like the Higgs boson.

An interesting element in my model is that it can predict the existence of an as-yet-undetected fourth generation of quarks. I figured that the *effective* constituent mass of the 125- to 126-GeV boson should be approximately twice the mass of a 63-GeV quark. Until now, such a fourth-generation quark does not appear to have been observed, because no quark has been seen between the bottom quark, with a mass of about 4.5 GeV, and the top quark, with a mass of 173 GeV. (The other quarks are all lower than 4.5 GeV in mass.)

Because a quark with a mass of about 63 GeV has not yet been detected, I postulated that the zeta and zeta prime boson masses are determined by the combination of the masses of the bottomonium and toponium bound states. There is a formula from the mixing of the zeta and zeta prime resonances such that when you insert the masses of the bottomonium and toponium states, and identify the zeta with the 125-GeV boson, then the mass of the zeta prime is 230 GeV. The "mixing" of these states means they interact with one another in a particular way, because the bottomonium and toponium states have identical quantum numbers, although different masses. The same is true of the zeta and zeta prime mesons. Similar mixing of bosons has already been confirmed experimentally for the neutral K-mesons and their anti-K-mesons, and also for the mixing of neutrino flavors, such as the electron neutrino and the muon neutrino, which resolve the solar neutrino problem of not enough electron neutrinos being emitted from the sun. In addition, there is strong evidence for the mixing of the pseudoscalar eta and eta-prime mesons, which belong to the lower mass octet of pseudoscalar mesons. A problem with the assumption that the zeta and zeta-prime resonances are formed from a mixing of the bottomonium and toponium states is that the standard gluon forces in perturbative QCD are not strong enough to cause the mixing that will form the zeta and zeta prime. It may be that a new strong gluon force has to be assumed to be acting between top and antitop, and bottom and antibottom, quark states.

This idea of mine differed from the standard quark flavor mixing, which is determined by what is called the *Cabibbo* and the *Cabibbo–Kobayashi–Maskawa matrices*, which mix the flavors of the six known quarks and leptons within the standard model. In my model, I, of course, keep these standard quark flavor mixes.

My Higgs impostor particle, the zeta, is electrically neutral and is analogous to the well-known parapositronium particle that can decay into two photons. Otherwise, the rest of the model is based on standard QCD and quark–gluon interactions, enhanced by nonperturbative gluon interactions.

The ability to produce a theoretical Higgs impostor such as the zeta underscores the critical need to determine the parity and spin of the new 125-GeV boson. If the experimentalists eventually find the new particle to have spin 0 and positive parity, then they will have found the Higgs boson, whereas if they identify the new boson as having spin 0 but negative parity, this indicates a pseudoscalar boson, and they will have possibly found the zeta particle. Other impostor models have been proposed. For example, Estia Eichten, Kenneth Lane, and Adam Martin have proposed that the 125-GeV boson be identified as a Techni-boson in a modified Technicolor model. (Eichten and Lane are members of the Fermilab group in Chicago, whereas Martin is a member of CERN.)

The other way to distinguish the zeta and zeta-prime bosons from the Higgs boson is to analyze carefully the rates of their decays into lower mass particles. The decay rates of the zeta are predicted by quarkonium calculations in QCD, whereas the Higgs boson decay rates are determined by the standard electroweak theory, given a Higgs mass of 125 GeV. The decay rates of the two bosons, the zeta and Higgs, into fermion–antifermion pairs, such as bottom and antibottom quarks, will be significantly different.

AUGUST 7, 2012

The Perimeter Institute ran a conference titled LHC Search Strategies from the 2nd to the 4th of August, which I attended. There were more than 40 participants from different international laboratories and universities, equally divided between theorists and experimentalists. The purpose of the workshop was to discuss what experiments should be done next to investigate the properties of the new boson. Proposals to discover physics beyond the standard model, such as supersymmetric particles and other possible exotic particles, played a dominant role in the workshop.

The first morning started with introductory talks by the two CMS group leaders, Joe Incandela and Greg Landsberg. Incandela, from the University of California at Santa Barbara, gave a clear presentation about how the CMS and ATLAS groups claim to have discovered a new particle at 125 to 126 GeV. As in the seminar he presented on July 4 at CERN, he was cautious about claiming that they had found the standard-model Higgs boson.

Next, Landsberg, from Brown University, discussed the status of the CMS and ATLAS results. He stated that identifying the spin and parity of the new boson was a primary objective before the LHC closed down on February 17, 2013, for about two years. He claimed that by the end of 2012, they would have about 30 inverse femtobarns of integrated luminosity at an energy of 8 TeV, allowing them to determine the parity of the new boson to a sensitivity of 3 sigma. He also said that the decision regarding whether the spin of the boson is 0 or 2 would not be as easy to determine as the parity.

Landsberg claimed that it is difficult to fit a Higgs that is a spin-0 pseudoscalar boson into the standard model as an elementary particle. One possibility is to have two Higgs bosons, with the pseudoscalar nature of the Higgs boson revealing itself through quantum loop corrections. In my opinion, this is a contrived model and unlikely to be correct. Another possibility that Landsberg discussed is that supersymmetry is discovered at the LHC. Then, in the MSSM there are three neutral Higgs bosons and two charged Higgs bosons. Two of the neutral Higgs bosons have positive parity and one has negative parity, and therefore is a pseudoscalar boson.

At the coffee break, I talked to Joe Incandela and stressed that, before it is claimed that the new boson is a standard-model Higgs, it is critical to determine its parity, which will indicate whether it is scalar or pseudoscalar. He agreed, and repeated Landsberg's claim that they expect to have enough data at the end of the year to determine the parity of the new boson to within 3 sigma or standard deviations. He did say they had a result already that favored a scalar boson. However, the result was only 1 sigma or less in statistical sensitivity, and therefore it was obviously inconclusive.

Near the end of the coffee break, I talked to Greg Landsberg. Again I stressed the importance of determining the parity of the new boson. He repeated his assertion that they would know the parity within three standard deviations by the end of the year when they reach a luminosity of 30 inverse femtobarns. He did say that it would not be easy for them to determine the spin of the boson using the data accumulated by the end of the year. We agreed that the most likely spin of the boson would be spin 0, because the boson is observed to decay into two photons. The alternative—that it is a spin-2 boson with positive parity, which can decay into two photons— was not a likely interpretation of the new boson because it would not be easy to support this by the data. I told him I was happy to see experimentalists at CERN were being cautious about claiming the new boson is the standard-model Higgs. I said this did not seem to be the case with the majority of particle theorists, who seem to have decided it is the Higgs boson, without further experimental investigation of its properties. He laughed, and said, "The theorists already decided it was the Higgs boson before we had discovered anything!"

After the coffee break, there was another introductory talk by Marcus Klute, an experimentalist from the Massachusetts Institute of Technology (MIT). The first slide of his talk showed a Quotation from the director-general of CERN, Rolf-Dieter Heuer, at the July 4 announcement: "We have it!" Klute added to the quote: "We do?" He is a member of the CMS group that discovered the new boson in the diphoton decay channel. He reviewed in detail the experimental findings related to the two golden channels. Klute's first slide indicated again that the CERN experimentalists were being cautious about the claim that the standard-model Higgs boson had been discovered.

Raman Sundrum, a theorist from Johns Hopkins University, spoke next. He and Lisa Randall, from Harvard, are famous for inventing the Randall–Sundrum model, which attempts to resolve the weak interaction energy scale and Planck energy scale hierarchy problem. They introduced the idea that the universe consists of two four-dimensional "branes" that are connected by a five-dimensional "bulk." These branes are part of the brane extension of superstring theory in which not only are there two-dimensional strings, but also there are higher dimensional branes. The standard model of particle physics inhabits one of these branes. The five-dimensional spacetime warps the four-dimensional spacetime in such a way that it reduces the tension between the energy of the electroweak scale at about 250 GeV and the Planck scale at 10^{19} GeV. This model has been studied extensively by the particle physics and string theory communities. It predicts the Kaluza–Klein particles in the five-dimensional scenario.

In his talk, Sundrum began by saying that, now that a new boson has been discovered at 125 GeV, which is consistent with the standard-model Higgs boson, we are left with serious "naturalness" problems—namely, the Higgs mass hierarchy problem, the severe fine-tuning of the vacuum energy density, and the electroweak energy scale and Planck scale fine-tuning problems. He said that these problems were painful for theorists to endure, and put his hand on his stomach as a gesture of pain.

The naturalness problem has been part of the history of the standard model since the inception of the idea that we need a Higgs boson to produce the masses of elementary particles and to make the theory renormalizable. Sundrum reviewed the usual attempts to resolve the Higgs mass hierarchy problem, including the MSSM, the Little Higgs solution, and composite models of the Higgs boson such as Technicolor. Theorists abhor any severe fine-tuning in their fundamental theories, and the fine-tuning problems he mentioned are quite severe and make the standard model unattractive unless a BSM scenario is discovered at the LHC. So far, the LHC has not discovered any hints of BSM physics (see Chapter 10 for a detailed discussion of the naturalness problem).

The rest of the workshop consisted mainly of discussion groups that took place in two adjacent seminar rooms at the Perimeter Institute, the Space Room and the Gravity Room. One of the discussion groups concentrated on experimental strategies for the LHC's search for new exotic particles such as supersymmetric particles, whereas the other group concerned itself with theoretical issues such as the naturalness problem and searching for BSM physics.

I attended several of the experimentalists' discussion sessions, and in one, which was organized by CERN experimentalist Albert De Roeck, I mentioned the importance of determining the spin and parity of the new boson, and discussed how this could be done by determining the angular distributions of the decay of the new boson into two Zs and their subsequent decay into four leptons. I emphasized that the spins of all the observed particles in the standard model were either spin 1 or spin ½. Because the new boson had been observed to decay into two photons, it could not have spin 1, because the two resulting photons each have spin 1; the new boson, then, must have either spin 0 or spin 2. I repeated Landsberg's statement in his talk that a 3-sigma separation between the scalar and pseudoscalar parity assignments of the new boson could be reached when the LHC had accumulated enough data by running the machine at 30 inverse femtobarns of luminosity at 8 TeV later in the year. I emphasized that no collider had ever detected an elementary boson with the quantum numbers of the vacuum.

De Roeck nodded in agreement after my little lecture, and said that experimentalists at the ATLAS and CMS detectors were using methods similar to what I had described to determine the spin and parity of the new boson. He agreed that this was an important experimental issue to resolve before they could conclude that the new boson was the standard-model Higgs boson.

At one of the theory sessions that I attended, a speaker had covered the blackboard with little t's, s's, b's, W's, Z's—all with tilde hats on them, signifying that they were "squarks," "winos," and "zinos," the supersymmetric partners of the observed standard-model particles. A popular way to solve the Higgs mass hierarchy problem has always been based on this supersymmetric model. However, the 2011 and 2012 data from the LHC have not detected any supersymmetric particles up to higher and higher energies.

At a lunch break I spoke to João Varela, who is a senior experimentalist at the LHC. He explained how amazing the computer analysis is of the data pouring out of the CMS and ATLAS detectors. The data are sent out to a large grid of computers around the globe. The analyzed data are then sent back to CERN and stored in "parking" facilities. There is so much data being produced by the trillions of proton–proton collisions that these "parking lots" have to be used to avoid an extreme traffic jam of data. The parked data are stored and sorted daily by mechanical CERN robots. After the final data are analyzed, the results of the scattering experiments are available immediately to CERN experimentalists in real time.

An important factor in guaranteeing the results obtained truly represent what is happening in nature is the use of "blind analysis." The data analysis goes through a very complicated software process involving statistical algorithms. Not until all the final analyses are completed is the conclusive result "opened."

I asked João whether there was any danger of the huge data storage and the resulting analysis being hacked by computer hackers. I said, "They've hacked into the Pentagon and large financial institutions, so why shouldn't they be able to hack into your system?"

He said yes, absolutely, attempts had been made to hack into the system but they were so far not successful. The way they avoid hacking is to isolate the data storage facilities at CERN from any outside influence.

The Perimeter Institute workshop indicated to me that the experimentalists connected with CERN were still proceeding cautiously, aiming to investigate the important attributes of the new boson before declaring it to be the standard-model Higgs. This is still in stark contrast to the media and physics blogs circus, where in many cases people have stated categorically that the Higgs boson has been discovered beyond doubt. For example, in his recent blog, "Not Even Wrong," Peter Woit presented his take on the discovery of the Higgs boson at the LHC. "Last month came an announcement from Geneva that physicists of my generation have been anxiously awaiting since our student days nearly forty years ago," he wrote, quoting from an article he had published in the Italian Left-wing newspaper *Il Manifesto*. "The Higgs particle showed up more or less exactly in the manner predicted by the so-called Standard Model.... We saw the equations of our textbooks dramatically confirmed."

Yet perhaps some theorists are modulating their initial certainty about what the LHC has discovered. In his blog, "Of Particular Significance," Matt Strassler, like Peter Woit, displayed a strong bias toward saying the Higgs boson had been discovered soon after the July 4 announcement. However, Strassler attended the workshop at the Perimeter Institute and wrote an article on his blog about it. He now talked about the "Higgs-like" boson, indicating that he had developed a more cautious approach to interpreting the new resonance.

SEPTEMBER 10, 2012

Stephen Hawking came to visit the Perimeter Institute for two weeks. This was after the inauguration of the new wing of the building, called the Stephen Hawking Centre. We were informed by our director, Neil Turok, that it was possible to meet Hawking and discuss research topics with him. He appeared at lunch with his nurse, seated as always in his supercomputer wheelchair. At lunch, I was joined by other Perimeter Institute colleagues and Jim Hartle, who had been invited from California to be present during Hawking's visit to conduct some collaborative research with him.

I recalled that during a sabbatical leave at Cambridge in 1972, I arrived a couple of times at the Department of Applied Mathematics and Theoretical

Physics in Silver Street and found Stephen sitting in his wheelchair, waiting for someone to turn up to help him up into the building, which I did. At that time, 40 years ago, when he was about 30, Stephen was in much better shape physically and was able to talk to some degree. He was often seen whizzing around Cambridge in his motorized chair, a danger to himself and others as he reconnoitered the Cambridge traffic. This was before he acquired his remarkable electronic wheelchair, which he uses today.

Stephen was now almost completely paralyzed. He had a metal contraption attached to his right cheek. By twitching a muscle in his cheek, he was able to communicate with his computer and select letters on his screen, thereby forming sentences, which his electronic voice synthesizer read out.

I sought out Stephen and found him in an office set aside for him and his nurse. Jim Hartle was present in the office, together with another collaborator, both of them armed with clipboards and paper. I wondered how such a collaboration could proceed by only communicating with Stephen through his very curtailed means of dialogue. He sat facing the door as I walked in. I said, "Stephen, I'm John Moffat. You remember we have met in the past." I thought I noted a glimmer of recognition in his eyes. I apologized to Jim Hartle for the interruption, explaining that I was leaving Waterloo the next day, and this was the only opportunity I had to see Stephen.

I approached Stephen and said that I wanted to discuss some research with him. I told him I was currently studying the properties of the new boson discovered at the LHC.

"Stephen, in my opinion it's too early to say that the new boson is the standard-model Higgs boson," I said. "For example, we do not yet know the spin or parity of the boson, and these are important properties that must be understood experimentally."

Now I waited while Stephen formed a response to my statement. I watched as he selected the letters on his screen to form his sentence. This took several minutes. Eventually, the sentence was complete on his computer screen, and as I read it, his electronic voice said, "I hope it is not the Higgs boson."

It was well known that Stephen had bet Gordon Kane, a professor at the University of Michigan at Ann Arbor, $100 that the LHC would not discover the standard-model Higgs boson. His opinion was based on work he had done on black holes. I said, "Stephen, it is too early for you to pay your $100 bet."

The nurse seated nearby leaned forward and said, "Stephen! You have already paid that $100!"

I waited while Stephen began to form another response. I leaned over his shoulder and watched the sentence forming on his screen. Eventually I read, as the electronic voice spoke, "If it is not the Higgs, I will claim back my money."

Then I said, "We have to wait for more data to determine important properties of the new boson. Hopefully, these data will be available by the end of the
year or early next year."

Then I waited again patiently while Stephen provided a longer response to
my latest statement. On the screen the words came slowly: "We need more data
and we must wait until they are available before we make a decision about the
Higgs."

I said, "Thank you, Stephen. Take care." I left, nodding my thanks to the
nurse. In the meantime, Jim Hartle and his collaborator had left the room and
were working nearby.

OCTOBER 31, 2012

Blog sites in particle physics can keep one up to date on rumors about new LHC
data and the general feeling in the physics community about what has been
discovered. In particular, after CERN's July 4 announcement of the discovery of
the new boson, there was euphoria. The blogs went all the way from demand-
ing that Peter Higgs and a choice of two others from the five living founders
of the standard Higgs boson model be awarded a Nobel Prize immediately to
cautionary statements that we did not know enough about the properties of the
new boson to be sure that it was the Higgs boson. In the United Kingdom, blogs
proclaimed that Peter Higgs should be knighted. On YouTube, Peter Higgs was
shown choking up and becoming teary-eyed as he sat in the audience listening
to the announcement of the new boson on July 4.

Since that announcement, a number of new papers have appeared on the
electronic archive explaining how the critical measurements of the parity of the
new boson could be performed at the LHC. To get a three-standard deviation
(3-sigma) separation in the data between the scalar (*positive*) parity and the
pseudoscalar (*negative*) parity distinction for the new boson would require a
significant luminosity and new data, which hopefully could be obtained before
the LHC shut down in February 2013. Many papers now referred to the new
boson as the "X particle," indicating that the authors were aware of the fact
that we were not yet in a position to claim the discovery of the standard-model
Higgs boson. The consensus among the theorists seemed to be that the new
boson is indeed the standard-model Higgs boson, whereas the consensus
among experimentalists was that it was a wait-and-see game, while we anticipate a more definitive statement about the identity of the new boson with the
advent of new data.

It was becoming apparent to me that if the new LHC data confirmed that
the new boson was a pseudoscalar particle, not a scalar elementary particle as

required by the standard model, then this would have serious consequences for the standard model. Some particle theorists at the Perimeter Institute and elsewhere claim that we already know that the new boson is a scalar particle because the standard model requires it to be so! I have told them that I defer to experimental physics and nature to make decisions about the future of particle physics, and try not to be persuaded by theoretical prejudices. I have made it clear to the particle theorists that if the data determine that it is a pseudo scalar boson, then the coupling of the pseudoscalar Higgs boson to the W and Z bosons would be significantly different from the coupling of the elementary scalar boson to those bosons. A consequence of this is that the standard model would no longer be a renormalizable theory.

It should be understood that the current analysis of the parity of the new boson is model dependent. The experimentalists compare the standard scalar Higgs boson model to an "effective" phenomenological pseudoscalar Higgs boson model. The latter model cannot be consistent with the current data unless it is manipulated artificially. A true analysis for parity should compare the standard Higgs boson model with an alternative non-Higgs boson model, such as my composite quarkonium resonance.

A problem with the pseudoscalar Higgs boson model is that its coupling to the two Z bosons would lead to a decay rate into the eventual four leptons that would be more than an order of magnitude smaller than the result yielded by the scalar Higgs boson. This would destroy the possibility of fitting the pseudoscalar Higgs decay into four leptons, which is one of the crucial golden decay channel results. This decay channel was critical in the CMS and ATLAS experiments, leading to the announcement of the discovery of the new boson. Future analysis of the data to determine the spin and parity of the new boson should eventually be able to avoid a model dependence.

The particle physics community is now waiting anxiously for new data to be "unblinded." After the data are returned to CERN by the worldwide grid of computers, the data are blinded, or locked up until about two weeks before a new announcement is made. The fact that the computer analysts around the world do not know the overall results is analogous to a blinded study in medical research.

In actuality, the data *are* unblinded to a few analysts a week or two before the official unblinding of the data to set the parameters of the algorithms used to obtain the final results. Therefore, strictly speaking, there is some bias built into the unblinding of the data, which leads to an important issue regarding the ultimate veracity of the result of the data analysis. There can be two outcomes of the blinding and unblinding of the data: the experimental groups accept the result as it is conveyed to them through the statistical algorithms or they do not accept the result and manipulate it to remove perceived errors. The latter action can lead to human bias in interpreting the results. For example, certain perceived "outliers" in the data can be removed, according to information shared by experimentalists online, which changes the final result significantly. The many CMS and ATLAS analysts have meetings in private to discuss the results of the data analysis before any official announcements are made. These meetings are not open to the public or to other physicists, including those at CERN, who are not part of the inner circle of analysts. We must have confidence that the analysts make every attempt to prevent bias in analyzing the data and in deciding what to announce.

JANUARY 2013

The next presentation of new data from CERN after the July 4 seminar and press conference took place at a high-energy physics conference at the University of Kyoto on November 14, 2012. The new results reported at the Kyoto meeting were not significantly different from the previous data. I followed the talks, posted on the conference website, with great interest. A significant fact that emerged from the Kyoto talks was that the CMS group had not updated its data for the decay of the new boson into two photons. This was the cause of several rumors appearing on blogs regarding the status of the CMS two-photon decay results. However, the latest combination of data from the two golden channels determines the mass of the new boson to be 125.8 GeV. This measurement of the mass is claimed to be accurate to within half a percent. In contrast, the ATLAS determination of the mass of the new boson using these two golden channels was 126 GeV, also with an error of about half a percent.

Moreover, it was rumored on the physics blogs and from supposed leaks from the CMS collaboration that the signal strength for the two-photon decay channel result had decreased from its original 4-sigma value announced at the July 4 CERN seminar. However, these were just rumors, and there was no confirmation of their validity from CERN in Kyoto.

Alexey Drozdetskiy, who belongs to the University of Florida, Gainesville, particle physics group, and who spoke on behalf of the CMS collaboration in Kyoto, presented the results for a spin–parity analysis. The analysis of the golden channel decay of what we are now calling the "X boson" into ZZ* and then into four leptons relied on sophisticated statistical methods called *multivariate analysis*, designed to determine the parity of the X boson. It compared the standard-model Higgs boson prediction for the angular correlations of the four leptons with the predictions of the effective pseudoscalar Higgs boson for the same decay. The effective pseudoscalar Higgs boson model did not follow naturally from the standard Higgs boson model, and it was not renormalizable.

In addition, we recall that this effective pseudoscalar Higgs model would only fit the ZZ* decay data if one manipulated its coupling strengths artificially by increasing the interaction strength more than 100 times. The Florida group claimed a 2.5-sigma result favoring a scalar spin-0 Higgs boson. The ATLAS collaboration also claimed the scalar positive parity is favored over the pseudo-scalar negative parity for the new boson.

Other talks given at the Kyoto meeting still showed that there was a deficit in the signal strength reported by both the CMS and ATLAS collaborations for tau+–tau– and bottom–antibottom decay channels of the X boson. In view of this, whether the X boson was indeed the much-sought standard-model Higgs boson is still an unresolved question.

Yet, theorists and experimentalists alike are now increasingly claiming that the overall data accumulated in the search for the Higgs boson at the LHC is consistent with a standard-model Higgs boson. Already, an institute is being formed and prizes are being awarded *as if* the new discovery is the Higgs boson. With the supposed dramatic discovery of the Higgs boson at CERN, Peter Higgs's name suddenly became universally famous throughout the world. The University of Edinburgh, where Peter is a retired professor, quickly capitalized on this and inaugurated the new Higgs Centre for Theoretical Physics in January 2013. On this occasion, the university held a Higgs Symposium attended by several well-known particle physicists. Among them was Joe Incandela, senior experimentalist in the CMS collaboration. He gave a detailed review of the CMS results in their search for the Higgs boson. These results did not contain anything essentially new beyond what had been announced at the Kyoto meeting.

In addition, Russian billionaire Yuri Borisovich Milner, an entrepreneur and venture capitalist, started the Fundamental Physics Prize Foundation in July 2012 and chose the first nine winners of the $3 million annual award. This prize is now the largest academic award in the world, beating both the Nobel and Templeton prizes. In addition to the nine winners, two $3 million special prizes were awarded in 2012 to Stephen Hawking and to seven scientists who led the effort to discover a Higgs-like particle at the LHC, including Joe Incandela and Fabiola Gianotti.

10

Do We Live in a Naturally Tuned Universe?

When constructing models of nature, theoretical physicists expect that calculating fundamental constants, such as the mass and charge of elementary particles, only involves arithmetical operations such as the subtraction of two real numbers that may include a few decimal places. Numbers like 2.134 are of order one, or about the size of one or unity, compared with 21,340,000, which is nowhere near the size of unity. To obtain a number of order one from the much larger number requires *fine-tuning*—in other words, subtracting a number of similar size. A theory that needs no fine-tuning when determining physical parameters, such as the mass and charge of the particles, is considered "natural." Theoretical physics contains paradigms, such as electromagnetism, QED, and classical general relativity, that have satisfied this naturalness criterion. If it is not satisfied—and the calculations in the theory involve canceling two numbers to many decimal places, rather than a few—then we can expect that the theory is flawed. The current electroweak theory in the standard model suffers from this lack of naturalness. A theory that has this fine-tuning problem is not falsified by this failing, but it signals strongly that the theory is incomplete and needs serious modification.

FINE-TUNING AND THE HIGGS MASS HIERARCHY PROBLEM

In the event that future data confirm beyond a doubt that the new boson discovered at the LHC at 125 GeV is the standard-model Higgs boson, and at the same time no other new particles or new physics are discovered, then we face a crisis in particle physics. When the Higgs mass is calculated from the standard model, a serious fine-tuning occurs resulting from the divergent or infinite nature of the calculation. This is called the *Higgs mass hierarchy problem*, which we encountered earlier in this book. The theoretical prediction for the Higgs mass comes from subtracting one potentially infinite term from another.

The scalar-field Higgs model is renormalizable, which means that infinities that occur in the calculations of the mass and charge of a particle can be absorbed by unobservable quantities, such as the bare mass and the self-interaction mass of the particle. This occurs during renormalization to yield a finite measured mass and charge. The renormalization involves infinities that grow significantly with increasing energy, to such an extent that it becomes difficult to control them in the renormalization scheme. Theories are not necessarily expected to be valid up to infinitely high energies. They have to be formulated with a built-in energy cutoff—the highest energy at which the theory remains valid. We cannot have an infinite quantity in the calculation of a constant, so we "cut off" the energy and thereby limit the size of the energy in the calculation. After a theory is defined with such a cutoff, the potentially divergent or infinite terms appear as powers of the cutoff. In the jargon of theoretical physicists, we then have *logarithmic*, *linear*, *quadratic*, and so on, *divergences*. These quantities, which are dependent on the cutoff, become infinite when the cutoff goes to infinity.

The least harmful of such infinities are those that involve the logarithm of the cutoff in energy needed to make sense of the renormalization calculation. The calculation of the quantum corrections to the Higgs boson mass gives a result proportional to the square of the energy cutoff, which increases drastically as we increase the energy, compared with the logarithm of the cutoff, which has a much slower increase in energy. If the standard model is valid all the way to the Planck mass–energy of 10^{19} GeV, then this leads to a fine-tuning of about one part in 10^{17} for a Higgs mass of 125 GeV! For this size of quantum correction, the bare mass of the Higgs has to have the miraculously negative value of this size, so that an improbable cancelation occurs to give the Higgs boson a mass of 125 GeV.

Observation tells us that the Higgs-like boson mass is about 125 GeV, but the calculated quantum energy in the electroweak theory makes the Higgs boson mass bigger. This quantum energy is the result of the interactions of the Higgs boson with virtual particles such as the top quark and the W and Z bosons. The quantum physics keeps the Higgs boson light only through a ridiculous degree of fine-tuning. In the perturbative scheme used to calculate the masses of the elementary particles, a difference occurs between the formally infinite "bare" mass, which is the particle's mass in the absence of interactions with other particles, and the infinite quantum energy mass. If the quantum energy behaves as a logarithm of the energy cutoff, then it can be controlled; it produces a result that does not require a fine-tuned cancelation between the unknown bare mass and the amount of quantum energy. On the other hand, if the quantum energy is the square or higher power of the cutoff, then the amount of quantum energy becomes uncontrollable, resulting in a fine-tuning to many decimal places in its cancelation with the bare mass to produce the experimentally observed mass.

We have an analogous situation in classical electromagnetism when an electron interacts with other electrons through the Coulomb force. The self-energy of the electron, which is produced by the electron's interaction with its own electromagnetic field, produces a large mass as a result of its large charge density. The contribution to the electron mass that comes from self-energy adds to the intrinsic bare mass of the electron (the mass that the electron would have in the absence of interaction). In the theory of classical electromagnetism, it is not difficult to see that the electron self-energy becomes infinite as the radius of the electron becomes zero. To get the small measured electron mass, a precise, fine-tuned cancelation must occur between the bare electron mass and the self-mass. However, another particle—the positively charged electron, or positron—comes to the rescue. Pairs of virtual electrons and positrons create a cloud around the negatively charged electron that smears out its charge and reduces the amount of electrostatic self-energy to a manageable size, thus avoiding an unacceptable fine-tuning in the calculation of the electron mass. Quantum physics resolves the fine-tuning problem for the electron mass.

THE GAUGE HIERARCHY PROBLEM

The Higgs mass hierarchy problem is related closely to the gauge hierarchy problem, which refers to the enormous difference between the electroweak energy scale or value of 250 GeV and the Planck energy scale of 10^{19} GeV. Based on experimental results, we know the symmetry breaking in the electroweak theory occurs at about 250 GeV. However, if the electroweak theory is valid up to the Planck energy, which is considered the ultimate energy reachable in particle physics, when gravity becomes as strong as the other three forces of nature, then one has to ask why there would be no other physically significant energy scale occurring between those two energy scales. At lower energies, the strong QCD force has an energy scale of 150 MeV, where the confinement of quarks occurs.

Without new physics, there is no explanation for this enormous energy gap between the electroweak and Planck energies, which seems very unnatural. You would expect something to occur between these two energy scales. If there are indeed no new particles between the electroweak energy scale and the Planck energy scale, then this constitutes a "desert" in particle physics.

THE SPECTRUM OF FERMION MASSES

A third hierarchy problem in the standard model has to do with the huge mass spectrum of quarks and leptons. The ratio of the electron mass of 0.5 MeV to the top quark mass of 173 GeV is 10^{-6}. The situation is even worse for the neutrino mass, which is about 0.2 eV; here, the ratio to the top quark mass is about 10^{-12}. The standard model does not provide an explanation of why there is such an enormous range in the mass spectrum of the quarks and leptons. After the spontaneous symmetry breaking of the electroweak theory, the Higgs boson is supposed to impart mass to the quarks and leptons, but the actual experimental values of their masses are adjusted by hand in an ad hoc manner. We need to have a theoretical calculation that predicts the experimental values of the quark and lepton masses, and that explains the large, unnatural mass difference between the neutrino or the electron and the top quark.

These three related hierarchy problems in the standard model—the Higgs mass hierarchy problem, the gauge hierarchy problem, and the fermion mass hierarchy problem—along with the cosmological constant problem, all exhibit extreme fine-tuning and are known collectively as the *naturalness problem*. The naturalness problem has been a constant thorn in the side of the standard model, and has been discussed contentiously for decades in the particle physics community.

TRYING TO SOLVE THE NATURALNESS PROBLEM WITH SUPERSYMMETRY

One of the first attempts to solve the naturalness problem in the standard model was supersymmetry, proposed during the 1970s. Here, as in our example of the calculation of the electron mass, other particles come to the rescue to solve the Higgs mass hierarchy problem. These are the superpartners that differ from the known standard-model particles by a half spin, producing a fermion partner for every boson particle and vice versa. Because of the nature of the supersymmetric particle charges, the large quantum self-energy contribution to the Higgs boson mass is neatly canceled. This solution to the Higgs mass hierarchy problem is only valid for an exact, rather than broken, supersymmetry and for superpartner masses that are not too far removed from the observed standard-model particle masses. When calculating the boson masses in supersymmetry, the electric charges of the bosons have the opposite signs to the fermion charges, so for each boson there is an oppositely charged superfermion with approximately the same mass, and a cancelation occurs, reducing the fine-tuning of the Higgs boson mass to an acceptable level.

Martinus Veltman published a paper in 1981,[1] in which he investigated the naturalness problem in the standard model, concentrating in particular on the Higgs mass hierarchy problem. Using supersymmetry, he formulated a mathematical condition involving the masses of the standard-model particles, which, when satisfied, would solve the naturalness problem.

Yet, this supersymmetry resolution of the Higgs mass hierarchy problem has more or less been invalidated, because the LHC has not found any supersymmetric partners such as stops and gluinos up to energy bounds of 1 to 2 TeV. Or, in supersymmetry-speak, the LHC has not found any "spartners" such as stops and gluinos, which are the supersymmetric partners of the top quark and the gluon, respectively. As a result of the lack of evidence of superpartners below 1 to 2 TeV, the attempts to solve the hierarchy problems by using supersymmetry models lead, again, to an unacceptable level of fine-tuning.

TRYING TO SOLVE THE NATURALNESS PROBLEM WITH THE MULTIVERSE

Because there is no evidence so far of supersymmetry, physicists have been searching for other possible solutions to the hierarchy and fine-tuning problems. A recent spate of papers by notable particle physicists claims that we must consider the multiverse model. Here, within the posited vast (and possibly infinite) number of different universes, our standard model of particle physics occurs in a finely tuned way in only one—the universe in which we live. In other words, this is just the way our particular universe is, so there's no sense worrying about it. This ad hoc "solution" to the standard-model hierarchy and fine-tuning problems has been widely criticized in the physics community. For one thing, because we cannot access any of the other universes in the multiverse, we can never observe what goes on in any universe except our own, with the consequence that we cannot verify or falsify any part of the multiverse paradigm.

1. M. Veltman, "The Infrared-Ultraviolet Connection," *Acta Physica Polonica*, B12, 437–457 (1981).

The multiverse is related to the anthropic principle, which holds that many of the laws of physics are based on very finely tuned constants. The idea of the anthropic principle was first applied to physics and cosmology by Robert Dicke of Princeton University during the late 1950s and early 1960s. He said that the age of the universe was just right—namely, the universe chose a "Goldilocks" age, which made it possible for humans to have evolved to observe it. If the universe were 10 times younger than its 13.7 billion years, then much of astrophysics and cosmology would not be valid as we observe it today. In such a young universe, there would not have been time to build up crucial elements for life such as carbon and oxygen by nuclear synthesis. If the universe were 10 times older than it is measured to be, then most stars would have burned out and become white dwarfs. Planetary systems such as ours would have become unstable a long time ago.

In 1973, cosmologist and astrophysicist Brandon Carter promoted the so-called *weak anthropic principle* at a symposium in Kraków honoring the 500th birthday of Copernicus. The Copernican principle, flowing from Copernicus's discovery that the earth was not the center of the universe, states that human beings do not occupy a privileged place in the universe. Carter used the anthropic principle in reaction to the Copernican principle. He said at the symposium, "Although our situation is not necessarily *central*, it is inevitably privileged to some extent." Carter disagreed with using the Copernican principle to justify the so-called perfect cosmological principle.

This principle was the basis of the steady-state theory of cosmology proposed by Fred Hoyle, Herman Bondi, and Thomas Gold in 1948, and it held that all the regions of space and time are identical. That is, the large-scale universe looks the same everywhere, the same as it always has and always will. The steady-state theory was falsified eventually in 1965 by the discovery of cosmic microwave background radiation. This discovery provided clear evidence that the universe has changed radically over time, as described by the Big Bang theory, and is not the same everywhere, neither in space nor in time. There is a weaker form of this idea called the *cosmological principle*, which states that we do not occupy a special position in space, and the universe changes its gross features with time, but not in space (in contrast to the perfect cosmological principle, in which the universe never changes in time or space). The weaker form of the Copernican principle is the basis of the standard model of cosmology.

A book originally published in 1986 by John Barrow and Frank Tipler[2] distinguished between the weak and strong anthropic principles. The *weak* anthropic principle states that life as we know it exists because we are here as observers

2. J.D. Barrow and F.J. Tipler, *The Anthropic Cosmological Principle*, Oxford Paperbacks (New York: Oxford University Press, 1988).

in our universe. Roger Penrose has described the weak anthropic principle as follows: "The argument can be used to explain why the conditions happen to be just right for the existence of (intelligent) life on the earth at the present time. For if they were not just right, then we should not have found ourselves to be here now, but somewhere else, at some other appropriate time."[3]

The *strong* anthropic principle says that we are here in the universe because many constants and laws of physics are finely tuned. For example, if the electric charge of the electron were a tiny bit different from its measured value, or if the mass difference between the proton and neutron were a bit different than its measured value, then the laws of atomic and nuclear physics would fail, and we could not exist as a life form in the universe. Evolutionary biologist Alfred Russell Wallace, who, independent of Charles Darwin, proposed the theory of evolution, anticipated the modern version of the anthropic principle back in 1904: "Such a vast and complex universe as that which we know exists around us, may have been absolutely required...in order to produce a world that should be precisely adapted in every detail for the orderly development of life culminating in man."[4]

According to the multiverse paradigm, which is related to the anthropic principle, we have to accept that this is simply the way the universe is, and not attempt to explain away or overcome the extreme fine-tuning that occurs in the standard model, such as the hierarchy problems related to the Higgs boson, because many of the universes in the multiverse would have the same problems. In other words, the multiverse "solves" the naturalness problem by not considering it a problem.

More than 400 years ago, when Johannes Kepler attempted to explain the relative distances of the five planets known in his time from the sun and from each other, he was attempting to do it scientifically. He used the five Platonic solids to demonstrate why the distances between the planets were what they were (Figure 10.1). In effect, he was aiming for a "natural" explanation for the sizes of the planetary orbits, and he assumed that the planets existed independently of human observers on earth. His Platonic solids model gave approximately the right answer to the relative sizes of the orbits. Yet, from a modern astronomical perspective, Kepler's use of the Platonic solids to explain the planetary orbital radii appears quaint.

3. Roger Penrose, *The Emperor's New Mind: Concerning Computers, Minds, and the Laws of Physics*, Vintage edition (New York: Oxford University Press, 1989), 560–561.

4. Alfred Russell Wallace, *Man's Place in the Universe: A Study of the Results of Scientific Research in Relation to the Unity or Plurality of Worlds*, 4th ed. (London: George Bell & Sons, 1904), 256–257.

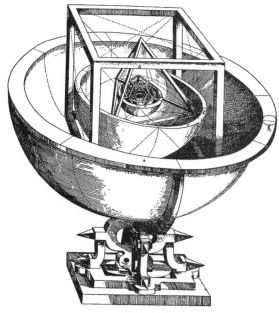

Figure 10.1 Kepler's five Platonic solids, a pleasing but wrong-headed idea that correlated the distances between the orbits of the known planets with the dimensions of nesting octahedron, icosahedron, dodecahedron, tetrahedron, and cube.
SOURCE: Wikipedia.

However, proponents of the anthropic principle today go beyond "quaint." They say that it was not even meaningful for Kepler to try to obtain an objective physical explanation of the relative distances of the planets. From their point of view, it was disappointing that Newton's gravity theory had failed to determine and constrain the radii of planetary orbits, but that was just the way things were in our particular universe. It made no sense to ask why.

Instead, they claim, the radii of the planetary orbits are what they are simply because of a historical accident during the formation of the solar system. Rather than focusing on the planets as objective and related celestial bodies, the anthropic principle proponents focus on the earth. They explain that the earth is just the right distance from the sun to create an environment for human beings to exist because, for one thing, the surface temperature allows for liquid water. The important point to them is that human beings developed on the earth, and the orbits of earth and the other planets contributed to that event. If we suppose that our planet is the only inhabited planet in the universe, their argument continues, then it would be miraculous for the earth to be located at exactly the right distance from the sun to support life. However, if we take into account the other planets in our solar system and the myriad other extrasolar system planets, then the "fine-tuning" of the earth's distance from the sun does not seem so unique. Many other planets in the universe might be as favorably situated.

In the anthropic view, the existence of human beings is *central*. Our existence explains away any concerns about puzzling planetary orbits or unwieldy constants or the necessity for arithmetical fine-tuning. We should just accept the conditions that allowed for our evolution.

Well-known scientists today are espousing the anthropic view, and science and the scientific method that we have known for centuries may be morphing into philosophy or mysticism. A natural universe is one in which there are no unnatural fine-tunings and there is no need to invoke the anthropic principle to explain the laws of nature. The basic question in considering the multiverse "solution" is: Should we continue to try to remove the fine-tuning problems in deriving the laws of physics—that is, should we continue to try to improve our theories about nature—or should we just accept these problems and accept that we live in an unnatural anthropic universe that might be part of a multiverse?

THE HIGGS BOSON AND THE NATURALNESS PROBLEM

Given the fine-tuning in the standard model and the proposed resolution by means of the multiverse and anthropic principle, we must ask the fundamental question: Do we live in a natural universe or not? If the new boson discovered at the LHC is, in fact, the standard-model Higgs boson responsible for the fine-tuning, the particle physics community cannot ignore the trouble this discovery brings with it. The trouble arises because of the failure to detect new physics beyond the standard model at the LHC, such as supersymmetric superpartners. Such new physics could alleviate the fine-tuning or naturalness problem. However, this requires a complicated conspiracy among any new particles such that the hierarchy problem gets canceled out. If the particles do not conspire to allow this cancelation, then, ironically, the situation could be even worse because the large masses of the new particles would increase the size of the quantum corrections responsible for the Higgs mass hierarchy problem.[5]

Is there a way, within the current formulation of particle physics, to resolve the hierarchy problems and produce a natural explanation of the standard

---

5. The term *quantum correction* is misleading because a correction contribution in perturbation theory is supposed to be smaller than the term that it is correcting. This may not be the case with a quantum energy "correction"; the correction can be actually larger than the initial classical value in the perturbation theory.

model, without new physics beyond the standard model, without recourse to severe fine-tuning, and without embracing the multiverse paradigm?

The source of the fine-tuning of the Higgs boson mass and the gauge hierarchy problem is the way the electroweak symmetry $SU(2) \times U(1)$ is broken in the standard model by a Higgs boson field. The Higgs field is introduced into the standard model to break the symmetry through a special potential, which contains a scalar field mass squared multiplied by the square of the scalar field. Added to this is the scalar field raised to the fourth power and then multiplied by a coupling constant, lambda. Choosing the scalar mass squared to be negative allows the potential as a function of the scalar field to have a maximum and two minima. The maximum occurs for the vanishing of the scalar field and the minima occur for constant, nonzero values of the scalar field (refer back to Figure 5.2). It is the minima in the ground state that break the basic $SU(2) \times U(1)$ symmetry and yield the positive physical Higgs mass (initially not including quantum mass corrections) as well as the W and Z boson masses.

Because the initial form of the Lagrangian in the electroweak theory has massless W and Z bosons, quarks and leptons, and a massless photon, the only constant in the Lagrangian with mass is the Higgs mass. This mass breaks the scale invariance or conformal symmetry of the classical Lagrangian. Theories that have massless particles such as photons satisfy a special symmetry invariance called *conformal invariance*. The original and best known theory satisfying conformal invariance symmetry is Maxwell's theory of electromagnetism, with massless photons traveling at the speed of light.

In 1980, Gerard 't Hooft proposed a resolution to the electroweak hierarchy problems in which the Higgs mass would be protected by a symmetry that would naturally make it light. In 1995, William Bardeen proposed, at a conference held at Fermi Lab, a way to avoid the hierarchy problems. To keep the initial classical Lagrangian of the electroweak model "conformally invariant," he set the Higgs mass in the potential to zero. This means that the original form of the mechanism for producing spontaneous breaking of the electroweak symmetry is lost. However, in 1973 Sydney Coleman and Eric Weinberg, both at Harvard, proposed that the symmetry breaking of the Weinberg–Salam electroweak model could be achieved by the purely quantum self-interaction of the Higgs boson.[6] These quantum self-energy contributions would produce an effective potential without a classical mass contribution in the electroweak Lagrangian. A minimization of this effective potential breaks the electroweak symmetry. If there are no new particles beyond the ones in the standard model that have been observed, then this could solve the hierarchy problems. In

6. S. Coleman and E. Weinberg, "Radiative Corrections as the Origin of Spontaneous Symmetry Breaking," *Physical Review*, D7, 1888–1910 (1973).

particular, the Higgs mass quantum contributions can be tamed to be small logarithmic corrections that can be accounted for without a fine-tuning of the Higgs mass calculation. Other authors, such as Krzysztof Meissner and Hermann Nicolai at the University of Warsaw, and Mikhail Shaposhnikov and collaborators at the École Polytechnique Fédérale de Lausanne, have developed these ideas further.

There is a serious problem with this potential solution to the hierarchy problems. The top quark mass is large (173 GeV), and it couples strongly to the Higgs boson in the postulated Yukawa Lagrangian, which is a part of the standard (Steven) Weinberg–Salam model introduced to give the quarks and leptons masses. Because of this coupling, the final calculation of the quantum field contributions to the Coleman–(Eric) Weinberg effective potential becomes unphysical, which illustrates how the standard model is tightly constrained and not easy to modify without invoking unwanted unphysical consequences.

Another source of fine-tuning is the neutrino masses. A massive fermion has both a left-handed (isospin doublet) component for the Dirac field operator as well as a right-handed (isospin singlet) component. However, the right-handed neutrino has never been observed. Because the theory demands its existence, particle physicists hypothesize that it must have a mass so large that it has not yet been detected at the LHC.

COSMOLOGICAL BEARINGS ON PARTICLE PHYSICS

The vexing problem of the cosmological constant—namely, having to explain why it is zero or, as is required observationally by the standard LambdaCDM cosmological model, why it has a tiny but nonzero value—only arises in particle physics when the standard-model particles couple to gravity. In the absence of gravitational interactions between the particles, there is no cosmological constant problem. The cosmological constant problem has currently not been solved in a convincing way and remains a puzzling issue that has to find a resolution before we can accept confidently the standard model of particle physics and cosmology as the final word. Recently, the Planck satellite mission collaboration published remarkably precise data revealing new features in early-universe cosmology.[7] In particular, these data demonstrate that the early universe is simpler than we had anticipated in early theoretical cosmological models. The new precise data eliminate many of the more complicated inflationary models such as hybrid models, which involve several scalar fields

7. P. Ade et al., "Planck 2013 Results. XXII. Constraints on inflation," arXiv.org/1303.5082.

with additional ad hoc free parameters. These new data bring into relief the long-standing naturalness problem: How can the universe begin with the Big Bang without extreme fine-tuning of its initial conditions?

The most famous model for solving the naturalness problem in the beginning of the universe is the inflation model, originally proposed by Alan Guth in 1981. The Planck data restrict the possible inflationary models to the simplest ones based on a single scalar field called the *inflaton*. In a recent paper, Anna Ijjas, Paul Steinhardt, and Abraham Loeb have pointed out that the simple inflationary model is in trouble.[8] Although a simple scalar inflaton model can fit the Planck cosmological data, these authors argue, the data make the inflationary model an unlikely paradigm.

It has been known for some years from the published papers of Andre Linde[9] at Stanford University, Alexander Vilenkin[10] at Tufts University, and Alan Guth at MIT[11] that the inflationary models must suffer "eternal inflation."[12] That is, once inflation is induced just after the Big Bang, it will continue producing a multiverse eternally in which, as Guth states, "anything that can happen will happen, and it will happen an infinite number of times."[13] The result is that all possible cosmological models can occur, rendering the inflationary paradigm totally unpredictive. If future observations continue to support the results of the Planck mission, then the inflationary paradigm may be doomed, which opens up the possibility of considering alternative cosmological models that can solve the initial value problems of the universe, such as the VSL models and periodic cyclical models.

In the first papers I published on VSL in 1992/1993, I raised criticisms of the inflationary models, which have been borne out by the latest Planck mission

8. A. Ijjas, P.J. Steinhardt, and A. Loeb, "Inflationary Paradigm in Trouble after Planck 2013," *Physics Letters*, B723, 261–266 (2013).

9. A. Linde, "Eternally Existing Self-Reproducing Chaotic Inflationary Universe," *Physics Letters*, B175, 395–400 (1986).

10. J. Garriga and A. Vilenkin, "Prediction and Explanation in the Multiverse," *Physical Review*, D77, 043526 (2008).

11. A. Guth, "Eternal Inflation and its Implications," *Journal of Physics*, A40, 6811–6826 (2007).

12. A consequence of eternal inflation is that quantum matter fluctuations that are the seeds of stars and galaxies become "runaway" and very large, and the universe becomes increasingly inhomogeneous as it expands, contradicting observation and the original motivation for introducing inflation models.

13. Guth made this statement to journalist Mike Martin of United Press International at a cosmology conference in Boston on March 23, 2001.

data. VSL models were also published in 1999 by Andreas Albrecht and João Maguiejo.[14] The periodic cyclic models, in which the universe expands and contracts with a new birth at the beginning of the Big Bang and a death at the Big Crunch, were considered by Richard Tolman during the 1930s. More recently, the cyclic model has been revived and published by Paul Steinhardt and Neil Turok in 2002.[15]

Trying to make the standard model of particle physics and cosmology natural without unacceptable fine-tuning problems has become a vital effort in particle physics today. It is possible that the LHC, at much higher energies after 2015, will not detect any new physics. Nevertheless, I believe that there does exist a resolution to the unwanted hierarchy problems in the standard model. It is just extremely challenging to find it. Further research must be done on how electroweak symmetry is broken and how particles get their masses. Historically, some problems in physics have proved to be especially hard to solve. Often, it takes decades to find a solution that is natural and does not resort to speculative, unverifiable, and unfalsifiable proposals such as the multiverse and the anthropic paradigms.

14. A. Albrecht and J. Magueijo, "Time Varying Speed of Light as a Solution to Cosmological Puzzles," *Physical Review*, D59, 043518 (1999).

15. P. Steinhardt and N. Turok, "Cosmic Evolution in a Cyclic Universe," *Physical Review*, D65, 126003 (2002). In both the VSL and cyclic alternative models, the quantum matter fluctuations do not exhibit runaway, uncontrolled growth leading to a multiverse scenario. Moreover, they predict that gravitational waves will not be observed as a relic of the Big Bang. Inflationary models predict that gravitational waves will be detected by experiments such as the Planck mission.

11

The Last Word until 2015

In February 2013, the LHC was shut down for two years of maintenance and upgrading to higher energy levels. The "last word" from CERN about the discovery of the new boson was delivered at the Rencontres de Moriond meeting March 2–16, 2013.

The Moriond workshops consisted of two consecutive sessions: the Electroweak Session and the QCD Session. Highly anticipated talks took place at both of these sessions, with representatives from the ATLAS and CMS collaborations giving updates on the latest analyses of the 2011 and 2012 data. On March 14, during the Moriond workshop, the CERN press office made the following seemingly unequivocal announcement: "New results indicate that the particle discovered at CERN is a Higgs boson." The hints of the discovery of the Higgs boson on July 4, 2012, followed by stronger hints at the Kyoto meeting in November 2012, had now culminated in the experimentalists' new analyses of the data. Increasing numbers of papers discussing the LHC data and the properties of the new boson since 2012 appeared to be accepting as fact that the LHC had indeed discovered the Higgs boson at a mass of approximately 125 GeV.

For example, a paper submitted to the electronic archive by a veteran CERN theorist in March 2013 stated: "Beyond any reasonable doubt, the H particle is a Higgs boson." (Here the "H particle" refers to the new particle discovered at the LHC; other physicists refer to it as the X boson.) Clearly, not only the popular media but the particle physics community has decided that the Higgs boson has been discovered. One can only assume that the Nobel Prize for Physics will be awarded to Peter Higgs and one or two other theorists in 2013 or possibly in 2014. Perhaps the Nobel committee will act conservatively, awarding two or three experimentalists at the LHC the prize in 2013 for discovering a new boson, and presenting the theorists with the prize the following year.

How convincing are these latest data and analyses?

THE TWO-PHOTON DECAY CHANNEL DATA

At the Moriond meeting, experimentalists presented new results updating the various decay rates, branching ratios, and signal strengths of the new boson's decay channels. In particular, the CMS collaboration finally updated their data for the two-photon decay channel for the X boson. This update, presented by Christophe Ochando, from Laboratory Leprince Ringuet in Palaiseau, France, during the QCD session, had been eagerly anticipated since the July 4 announcement; rumors had been swirling about why the CMS team had not updated these data in November.

It appeared that the team had had problems analyzing the 2011 data and accounting for an unexpected excess in signal strength of the two-photon decay when compared with the theoretical prediction of the Higgs boson decay into two photons. (See Figure 11.1 for a candidate event.) In 2011, the energy in the accelerator was 7 TeV with a luminosity of about 5 inverse femtobarns. As Ochando noted, based on the new 2012 data, the statistical significance of the two-photon decay signal had decreased since the July 4 announcement. The CMS results for the two-photon decay channel gave for the new 8 TeV data, with a luminosity of about 19 inverse femtobarns, a signal strength of 0.93 +0.34 or −0.32 for one method of analysis. Another method of analysis of the same data produced a significant decrease in the signal strength of the diphoton channel. The signal strength for this analysis gave 0.55 +0.29 or −0.27. This is a drop in the signal strength to 2 sigma compared with the 4-sigma result originally announced on the fourth of July.

What could be done about these conflicting data, for which it appeared that an excess in signal strength had now become a deficit? Would it help to combine the results over the two years of data gathering? There are two ways of interpreting the results of the experiments. One way is to determine the signal strength of the decay products by multiplying the production cross-section of Higgs bosons with the branching ratio of the decay into two photons; the more statistically significant the signal strength, the more likely that the new boson is the Higgs boson. The other way to interpret the results of the experiments is to determine the ratio of the observed cross-section to the cross-section predicted by the standard model; ideally, this ratio should equal unity (one), with a strong statistical certainty.

Ochando revealed that when the CMS collaboration combined the 7-TeV and 8-TeV data from 2011 and 2012, respectively, this resulted in the ratio of the observed cross-section to the predicted standard-model cross-section being equal to 0.78 +0.28 or −0.26 for a Higgs boson mass of 125 GeV. This result is statistically consistent with the standard-model prediction of unity. On the other hand, a different analysis produced, again for the ratio of the observed

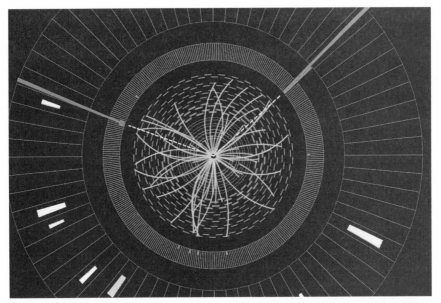

Figure 11.1 A candidate event for a Higgs boson decaying into two photons. © CERN for the benefit of the CMS Collaboration

cross-section to the standard-model predicted cross-section, the combined two-year result of 1.11 +0.32 or –0.30, a result also consistent with the standard model within experimental error. Combining the 2011 data and the 2012 data would also produce a signal strength with a larger statistical standard deviation consistent with the standard-model prediction.

What do these numbers mean? Are the CMS analysts correct in combining the results from two years? A couple of things appear to be clear. If the July 4 announcement had been based on just the new 2012 8-TeV data and a luminosity of 19 inverse femtobarns, then the 2- or 3-sigma strength for the diphoton decay channel would not have justified CERN announcing the discovery of a new boson! However, combining the 7- and 8-TeV data from the two years yields a result that strengthens the case for the standard-model Higgs boson.

Fabrice Hubaut from Aix-Marseille University presented the ATLAS results for the two-photon channel at the Electroweak Session. He announced the Higgs boson mass as 126.8 GeV, and the best-fit signal strength as 1.65 +0.34 or –0.30. This was a small decrease in the signal strength compared with the ATLAS data obtained at 7 TeV energy and 5 inverse femtobarns in 2011. The probability value for this result was a significant 7.4 sigma, which means that the ATLAS results still show a significant excess of events over and above the signal strength predicted by the Higgs boson model for the two-photon decay channel. The ATLAS results differ significantly from the CMS results, and the experimentalists will have to determine which result is correct.

Theoretical particle physicists have been discussing heatedly this excess of events in the two-photon decay, both in the earlier 2011 CMS data and even more so in the ATLAS data. One wonders how this large discrepancy between the new CMS and ATLAS data could have convinced unbiased physicists that the Higgs boson has been discovered "beyond any reasonable doubt."

Personally, I find it disturbing that the data for the two-photon decay are jumping about, and have not settled down to a statistically significant and definite value for both the CMS and ATLAS collaboration results. As for the reasons for the discrepancy, many papers have been published recently suggesting, for example, that the excess of events in the two-photon decay channel can be explained by including new charged vector boson particles in the calculation of the boson decay into two photons. If the excess of events in the two-photon decay is real, such charged particles would have to be detected at the LHC as new particles. Another explanation for the difference in the CMS and ATLAS results is that perhaps one detector could be more accurate in its ability to detect particles than the other. Some claim that the CMS muon–lepton detector is more accurate than the ATLAS detector. On the other hand, the noisy background problems and the differences in the methods of analyzing the data are complex and can certainly produce significant differences in the results. Perhaps it would be prudent to wait until these difficulties have been resolved before declaring the definite discovery of the Higgs boson.

OTHER DECAY CHANNELS

The other golden channel reported at the Moriond meeting by both the CMS and ATLAS collaborations, the X or H boson decaying into two Z bosons, which in turn decay into four leptons, showed an increase in signal strength since November 2012. The number of events observed above background had increased to a level that proved convincingly that a new boson had been discovered at an energy of about 125 GeV.

The CMS and ATLAS groups also reported new results for the important decay of the X boson into a pair of W bosons, which then decay into two neutrinos and two leptons. The results for the two detectors were different, with one claiming a stronger signal than the other. The problem with this decay channel is that, of the final four leptons, the two neutrino leptons are electrically neutral, and their presence can only be inferred through an amount of missing energy when one reconstructs the decay events and demands the conservation of energy and momentum. The need to determine accurately the missing neutrino energy causes significant background problems, because many particles are decaying into neutrinos as the collisions of the two streams of protons spew out debris. The distribution of the new boson decay events in this channel is flat, rather than appearing as a resonance peak. The calculation of the WW and neutrino background is very complex, and must be done carefully so as not to obscure the true boson signal.

A critical result reported at Moriond was the decay of the X boson into fermion–antifermion pairs, such as the leptonic tau⁺–tau⁻ decay channel and the quark bottom and antibottom decay channel. It is important to establish beyond a doubt that these decay channels are observed, because an important part of the standard electroweak model is that the Higgs boson couples to (decays into) fermions through the Yukawa interaction. The Yukawa Lagrangian in the standard model couples the Higgs boson to pairs of fermions. When one adjusts the coupling constants that measure the strengths of the interactions of the Higgs field, and substitutes the constant vacuum expectation value for the Higgs field, the masses of the fermions are produced. Without establishing firmly the coupling of the X boson to fermions such as quarks and leptons, it cannot be claimed that the newly discovered boson is the standard-model Higgs boson.

Victoria Martin from the University of Edinburgh reported the results of the ATLAS collaboration for the fermion–antifermion decay channels. She reported the signal strength for the tau⁺–tau⁻ decay as 0.7 +0.7 or −0.7, which can be considered a null result, consistent with the standard model, but also consistent with zero—that is, indicating no interaction of the Higgs boson with the tau leptons. She also reported null results for the bottom–antibottom decay channel, which is disappointing, because this is the *dominant* decay channel for a standard-model Higgs boson.

As Martin stated in her talk, the big problem in the tau–tau channel is the background caused by the huge number of decays of the Z bosons into pairs of tau leptons. This large background must be subtracted from the total detected signal to reveal the tiny signal of events resulting from the Higgs boson decaying into pairs of tau leptons. In practice, when reconstructing the tau lepton decay channel, one does not actually observe the tau leptons, because they have a short lifetime of 10^{-13} seconds. What one observes is the final decay products of the tau leptons: neutrinos, electrons, positrons, and muons. Again, because neutrinos are neutral, and because of the conservation of energy and momentum in the scattering process, the presence of the neutrinos is inferred from a missing energy in the scattering. The neutrinos contribute to a large, problematic background that has to be removed.

For the dominant bottom–antibottom decay channel, the background problem is a serious obstacle to determining the signal strength for this decay. The number of hadrons produced during the proton–proton collisions is enormous; indeed, the quark background is more than a million times bigger than the predicted signal of the standard Higgs boson decaying into bottom and antibottom quarks. Martin summed up the whole problem as "looking for a needle in a haystack."

Valentina Dutta of MIT reported on the CMS results for the same decay of the new boson into fermion–antifermion pairs. As with the ATLAS collaboration, the big problem at the CMS detector is cutting through the background to see the decay of the new boson into a pair of taus and bottom and antibottom quarks. But the CMS group handled the background produced by the reconstructed neutrino and lepton data differently than the ATLAS group for the tau pair decay channel. The reported signal strength for a pair of tau decays was 1.11 +0.4 or −0.4, or a little more than a 3-sigma signal strength, which is quite different from the null result reported by the ATLAS collaboration.

The difference in the CMS and ATLAS results for this fermion–antifermion decay of the new boson could be the result of systematic errors caused by instrument problems in one or both of the two detectors, but in my opinion this is unlikely. The difference is more likely a result of the calculation of the background. An error in the analysis of a few percent when subtracting the background to reveal the signal for the tau pair decay can cause a significant difference in the reported signal strength between the two collaborations. The same applies to the more problematic bottom and antibottom decay channel.

Which result should we believe for the critical decay of the new boson into a pair of fermions and antifermions? Unfortunately, unless both collaborations find new results by reanalyzing the 2012 data for these decay channels, we will have to wait for the startup of the LHC in 2015 at an energy of 13 TeV and a much higher luminosity of, say, 100 to 200 inverse femtobarns.

With no conclusive, current information about the two dominant decay channels (the bottom-antibottom and tau–tau), how can so many physicists, as well as the CERN press office, be so confident that the beast that may have been cornered is indeed the standard-model Higgs boson? Yet, as we can see in Figures 11.2 and 11.3, the combined 2011/2012 data for the important decay channels of the Higgs boson are consistent, within the current accuracy of the data, with the predictions of the standard model. These data represent, in 2011, an energy of 7 TeV and in, 2012, an energy of 8 TeV. We must conclude from Figures 11.2 and 11.3 that the *overall accumulated* evidence at this stage for the new boson being the standard-model Higgs boson has reached a statistically convincing level based on the decay channel data.

With the startup of the LHC in 2015 at an energy of 13 TeV and a luminosity of possibly 100 inverse femtobarns or more, the comparison of the new data with the standard model will most likely confirm the new boson *is* the Higgs boson.

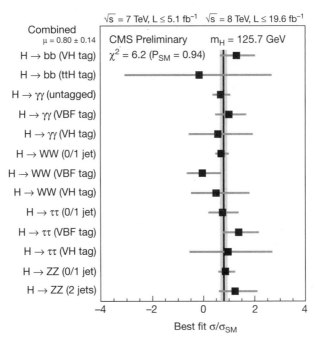

Figure 11.2 The best fit for the major Higgs boson decay channel data for the combined 7 TeV and 8 TeV data from the CMS detector. For 7 TeV, the luminosity is 5.1 inverse femtobarns; for 8 TeV, the luminosity is 19.6 inverse femtobarns. The best fit for the ratio of the observed cross-section to the standard-model predicted cross-section should be equal to one or unity. The horizontal bars represent the sizes of the errors deviating from the horizontal axis value of unity. The corresponding signal strength, or mu (μ), of these results is 0.80 ± 0.14. The signal strength of unity (one) would be a perfect fit to the standard model. © CERN for the benefit of the CMS Collaboration

DETERMINING THE SPIN AND PARITY

As we know, two important quantum numbers needed to identify a new particle are its spin and parity. The spin and parity of elementary particles are like items on ID bracelets, crucial to the identification of particles. As mentioned, spin is the intrinsic angular momentum degree of freedom of an elementary particle; parity is the operation on the particle of space inversion. That is, the x, y, and z coordinates associated with the particle are transformed to –x, –y, and –z. If, under the space inversion, the particle does not change sign, then it has positive parity and is a scalar particle; if it does change sign, it has negative parity and is a pseudoscalar particle. In practice, spin and parity are often considered together.

New results were presented at the Moriond sessions on the critical issue of the spin and parity of the new boson. As we recall, according to the theorem by Landau and Yang, the observation that the new boson decays into two photons

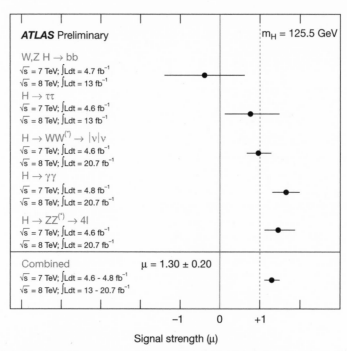

Figure 11.3 The ATLAS summary of the combined 2011/2012 data. Similar to the CMS data, the 2011 energy is 7 TeV with a luminosity of 4.6 to 4.8 inverse femtobarns, whereas the 2012 energy is 8 TeV with a luminosity of 13 to 20.7 inverse femtobarns. The horizontal axis represents the signal strength, mu (μ), and for the combined fit for the 7-TeV and 8-TeV data, μ equals 1.30 ± 0.20. A signal strength of unity (one) would be a perfect fit to the standard-model prediction. © CERN for the benefit of the ATLAS Collaboration

means that, because of conservation of angular momentum and spin, the boson can only have spin 0 or spin 2, and not spin 1, like the photons it decays into. So which is it? Spin 0 or 2?

Here we can turn to the other decay channels for information. The complicated analysis of the decay of the X boson into two Zs, which subsequently decay into four leptons (pairs of e+, e− and mu+, mu−), should be performed ideally by measuring certain angles to the horizontal axis on which the decaying particle sits at rest. Only five of the angles shown in Figure 11.4 are relevant for the determination of the spin and parity of the particle. The spin and parity are determined by a computer code that calculates the correlations of the measurements of the relevant angles.

A recent publication by the CMS collaboration[1] claims that the spin-0 scalar Higgs boson is favored by the data—and the scalar nature of the boson means

1. CMS Collaboration, "Study of the Mass and Spin-Parity of the Higgs Boson Candidate via its Decay to Z Boson Pairs," *Physical Review Letters*, 110, 081803 (2013).

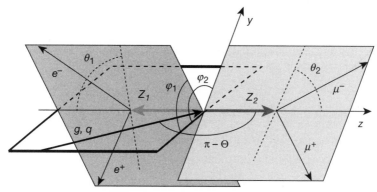

Figure 11.4 The Cabibbo–Maksymowicz angles in the H to ZZ decays. © "Higgs look-alikes at the LHC," by Alavaro De Rujula, Joseph Lykken, Maurizio Pierini, Christopher Rogan, and Maria Spiropulu. ArXiv:1001.5300v3 (hep-ph) 5 Mar 2010

that it has positive parity. The speakers from the ATLAS and CMS collaborations at the Moriond conference echoed this claim for the spin and parity of the X boson.

JULY 2013: FURTHER SPIN-PARITY ANALYSIS OF THE DATA

After the March 2013 Moriond workshop, the LHC collaborations continued their efforts to analyze the data that were collected in 2011 and 2012. An important new result is the determination of the spin and parity of the observed Higgs-like particle by the ATLAS group.[2] The group analyzed the decays of the X particle into two photons, into two Z-bosons further decaying into four charged leptons, and into two W-bosons further decaying into two charged lepton-neutrino pairs. (Recall that the first two of these are considered "golden" channels because they do not possess neutrino and hadronic background problems.)

The new analysis focusing on spin and parity uses a larger data set[3] collected by the ATLAS collaboration. The interactions of hypothetical spin-0, spin-1, and spin-2 resonances with standard-model particles are described in a way that

2. ATLAS Collaboration, "Evidence for the Spin-0 Nature of the Higgs Boson Using ATLAS Data," *Physics Letters* B 726, 120 (2013).

3. The integrated luminosity reached 20.7 fb⁻¹ at the energy of 8 TeV, with an additional 4.6 fb⁻¹ at 7 TeV.

depends on the theoretical model. The study focuses on specific production and decay processes that are the most relevant for spin and parity determination.

The statistical analysis aimed at determining the spin and parity of the X particle must be able to distinguish between signal and noise events. The specific statistical method, called *likelihood ratio analysis*, compares the likelihood of competing models being correct by comparing them to actual observations. One of the models tested is the standard-model Higgs boson, whose properties are known from the predictions of the standard model. Alternative models, such as the pseudoscalar Higgs model or the composite resonance model, are less specific in their predictions. However, their unknown properties can be described by what experimental analysts call "nuisance parameters." This way, comparing the standard-model Higgs boson with competing models can be independent of the assumptions of any specific competing model.

The likelihood ratio analysis requires knowledge of the statistical distributions of expected signal and noise events in both the standard-model Higgs case and competing models. The complex mathematics that describes the models makes it impossible to calculate these distributions using simple formulas. Instead, physicists use the so-called *Monte Carlo method*, which simulates a large number of events to obtain the required probability distribution for each competing model.[4] For each of the models, comparing actual observations to these simulated statistical distributions yields the likelihood value. These likelihoods can be compared, and their ratios calculated. Such ratios then determine how two competing models fare in the light of observations.

To determine the spin and parity of the observed Higgs-like particle, a favored process to investigate is the decay into two Z bosons and then into four leptons. Experimental physicists can measure a set of five distinct decay angles (Figure 11.4). The presence of neutrinos in the decay process that involves two W bosons complicates the analysis, as direct calculation of the decay angles becomes impossible. Unfortunately, no simple, conventional statistical method of analysis can discriminate between the Monte Carlo simulated distributions of competing models using presently available data. The ATLAS experiment, for example, produced 43 candidate signal events, 18 of which represent a signal for a 125.5-GeV Higgs boson.

There is a sophisticated statistical method that can help overcome these difficulties and improve the results of the likelihood analysis. This method is called

---

4. A good example of the Monte Carlo method entails the use of a simulated roulette wheel that can be spun a large number of times. The recorded outcomes of the ball landing on a particular number will eventually approximate accurately the theoretically expected probability distribution of the final result. Indeed, this method of statistical analysis was named after the famous Monte Carlo casino.

the *boosted decision tree algorithm* (BDT).[5] This is a learning algorithm that can be "trained" using simulated signal events to achieve maximum sensitivity so that it will be able to determine the probability distributions for the competing models. Using the likelihood analysis based on the BDT results, LHC analysts have shown that the data favor the standard-model prediction of a scalar, positive-parity Higgs boson over the alternatives. (See citations in footnotes 1 and 2.)

The ATLAS data for the decay of the X resonance into four leptons via two Z-bosons exclude the negative parity (pseudoscalar) hypothesis with a probability of 97.8 percent. The data for the decay into two charged lepton-neutrino pairs via a pair of W-bosons also agree with the standard model. These data can be used to exclude the spin-2 hypothesis for the new boson, too, but with less statistical significance.

The CMS collaboration's published paper, which obtained spin-parity results using statistical methods similar to those of ATLAS, also claims that the scalar Higgs spin-0 boson hypothesis is preferred over the pseudoscalar alternative.

The ATLAS collaboration excluded the alternative models that they studied without making assumptions about the strength of the couplings between the hypothetical Higgs-like particle and the other particles of the standard model. The accumulation of more data at energies 13–14 TeV with a much bigger luminosity will strengthen the statistical determination of the spin and parity assignments of the new resonance. If, during future spin-parity studies by both the CMS and the ATLAS collaborations, the statistical likelihood of the scalar spin-0 Higgs boson model continues to increase in comparison with the likelihood of alternative models, then we must conclude that the new boson with a mass 125–126 GeV is indeed the Higgs-like boson. Determining whether it is in fact the minimal, standard-model, elementary Higgs boson, or a more complicated particle, remains an open question.

LOOKING AHEAD

The standard-model Higgs boson, which is an electrically neutral scalar boson with spin 0 and positive parity, has the quantum numbers of the vacuum. Such an elementary particle has never been observed before. In the interactions of the elementary particles of the standard model, such as the quarks, the W and Z bosons, and the gluons and photons, the scalar Higgs boson has a somewhat

5. Although relatively new, the BDT algorithm has been used successfully in the past decade in such diverse areas as optical character recognition, speech recognition, neural networks, and cancer detection.

specter-like quality. The confirmation beyond a doubt that the new boson is indeed the standard-model Higgs boson, would be an extraordinary discovery in the history of science, for the particle has unique features that have never been observed experimentally until now.

Although we can claim that the overall results at the LHC for all the decay channels, the signal strengths, and the spin and parity obtained as of July 2013 are compelling in their compatibility with the standard-model Higgs boson, there are still problematic uncertainties in some decay channels and differences between the ATLAS and CMS results. As we know, details matter in physics. In 1916, the experimental difference between Einstein's theory of general relativity and Newtonian gravity was the tiny effect of 43 arc seconds per century in the perihelion advance of the planet Mercury. This is only about one percent of the total perihelion advance observed. However, such a tiny difference resulted in a huge change in our understanding of what gravity is. Another example of the importance of small details is Kepler's discovery that he could not resolve a tiny effect in the astronomical data of Danish astronomer Tycho Brahe by using circular orbits for the planets. This forced him into having to accept that the orbits of the planets are elliptical and it heralded the downfall of Ptolemaic astronomy and the acceptance of the Copernican heliocentric paradigm.

If the data to come out of the upgraded LHC in the coming years continue to strengthen the case for the new boson being the long-sought standard-model Higgs boson, then this will represent an enormous achievement in the long history of particle physics, as experiment validates theory. It will also represent one of the greatest technological feats in history, and should guarantee continued funding for the LHC and its successors.

Yet the discovery will not signal the end of the struggle to decipher the particle code of nature. As we know, the standard-model Higgs boson brings with it difficult problems to be solved, such as the Higgs mass hierarchy problem and the gauge hierarchy problem. These are in addition to the long-standing cosmological constant problem and the puzzle of how to combine Einstein's general relativity theory with quantum mechanics. After the celebratory party welcoming the standard-model Higgs boson officially into the company of elementary particles, physicists must get back to work and solve these problems that come in on the wings of the Higgs boson. We hope that they will be solved by the time-honored method of working out theoretical ideas and then verifying them by experiment, rather than resorting to pseudo-science resolutions such as the multiverse. The new boson, whether it is the standard-model Higgs boson or a different kind of Higgs boson, will offer new challenges for future generations of physicists working toward a deeper understanding of the origins and structure of matter.

ACKNOWLEDGMENTS

I thank my wife, Patricia, for her enduring and superb help in editing the book. Her tireless efforts in preparing the manuscript and her ongoing support made the book become a reality. I thank my agent Chris Bucci at Anne McDermid & Associates, Toronto, for his support and encouragement and for finding the right publisher. I am indebted to my colleagues Martin Green, Viktor Toth, and Vincenzo Branchina for reading the manuscript at an early stage and making many valuable suggestions for improving the book. I thank Bob Holdom and Alvaro de Rujula for helpful discussions. I am grateful to my editor Jeremy Lewis at Oxford University Press for being interested in the topic of particle physics and for publishing the book; thanks also to Erik Hane, Cat Ohala, and Bharathy Surya Prakash, who have been helpful in the final stages of the book preparation. I thank the John Templeton Foundation for their generous financial support for much of the research that forms the basis of this book, and for the writing of the book itself. I also thank my home institute, the Perimeter Institute for Theoretical Physics, for their support, and for providing opportunities for the exchange of ideas, which is a major way that theoretical physics progresses.

abelian and nonabelian groups   named after the 19th-century Norwegian mathematician Niels Henrik Abel, abelian groups satisfy the commutative rule (the results of a mathematical operation do not depend on the order of the elements [for example, $(a \times b) = (b \times a)$], whereas nonabelian groups do not satisfy this rule; their elements do not commute

accelerator   a machine that propels charged particles in beams at high speeds using electromagnetic fields; also known as a collider

action   the integral over time of the Lagrangian function, but for an action pertaining to fields, it may be integrated over spatial variables as well; it has the physical dimensions of angular momentum

anode   in a cell, battery, or apparatus with an electrical current, the anode is the positively charged electrode, which attracts negatively charged ions, most often electrons

anthropic principle   the idea that our existence in the universe imposes constraints on its properties; an extreme version claims that we owe our existence to this principle

antimatter   particles with opposite electric charge from their matter counterparts; predicted by Paul Dirac in 1928; the first antimatter particles, discovered by Carl Anderson in 1932, were positrons, or positively charged electrons

ATLAS and CMS detectors   two of four huge detectors at the large hadron collider (LHC) that collect and measure the debris from the collisions of proton beams; the acronyms stand for "a toroidal LHC apparatus" and "compact muon solenoid"

bare mass   the mass of a particle in the absence of interactions with other particles and their fields

baryon   a subatomic particle composed of three quarks, such as the proton and neutron

beta decay   a type of radioactive decay in which a nucleus of an atom emits an electron or a positron, and the nucleus receives its optimal ratio of protons and neutrons; beta decay is mediated by the weak force

Big Bang theory   the theory that the universe began with a violent explosion of space-time, and that matter and energy originated from an infinitely small and dense point

black body   an idealized physical system that absorbs all the electromagnetic radiation that hits it, and emits radiation at all frequencies with 100 percent efficiency

boson   a particle with integer spin, such as photons, mesons, and gravitons, that carries the forces between fermions

bubble chamber a chamber containing a superheated liquid, such as liquid hydrogen, through which electrically charged particles move, producing bubbles that allow the particles to be tracked and detected; the bubble chamber was invented by Donald Glaser in 1952

cathode a negatively charged electrode through which a current flows out of an electrical device; see *anode*

cloud chamber a sealed container filled with supersaturated water or alcohol vapor that ionizes when a charged particle interacts with it; an ion acts as a condensation nucleus around which a mist forms, allowing the track of the particle to be detected

collider a kind of accelerator in which directed beams of particles may collide against a stationary target or as two beams colliding head-on; colliders allow beams of particles to accelerate to very high kinetic energies when impacting other particles

color charge a charge carried by quarks and gluons that explains how quarks are confined inside hadrons, such as the proton and neutron; by combining their three colors (red, green, and blue), the resulting hadrons show white as their color charge; color charge also explains how three spin-½ quarks can be in the same quantum state inside a proton without violating Pauli's exclusion principle; quarks exchange gluons and create a strong color force field that binds the quarks together, becoming stronger as the quarks get farther apart

confinement the phenomenon of colored, charged quarks not being directly observed because they do not occur as free particles; quarks are confined in hadrons (three quarks) and in mesons (one quark and one antiquark)

Cooper pairs (of electrons) two electrons bound together at low temperature, forming the basis of the 1957 Bardeen, Cooper, and Schrieffer (BCS) theory of low-temperature superconductivity; the three shared the 1972 Nobel Prize

cosmic microwave background (CMB) the first significant evidence for the Big Bang theory; initially found in 1964 and studied further by NASA teams in 1989 and the early 2000s, the cosmic microwave background is a smooth signature of microwaves everywhere in the sky, representing the "afterglow" of the Big Bang; infrared light produced about 400,000 years after the Big Bang had red-shifted through the stretching of spacetime during 14 billion years of expansion to the microwave part of the electromagnetic spectrum, revealing a great deal of information about the early universe

cosmic rays very high-energy particles originating outside the solar system and penetrating the earth's atmosphere; they are composed mainly of protons and atomic nuclei that produce showers of secondary particles; recent data from the Fermi space telescope show that cosmic rays originate primarily from the supernovae of massive stars

cosmic void a vast, empty space in the universe between matter filaments that contains very few or no galaxies

cosmological constant a mathematical term that Einstein inserted into his gravity field equations in 1917 to keep the universe static and eternal; although he later regretted this and called it his "biggest blunder," cosmologists today still use the cosmological constant, and some equate it with the mysterious dark energy

Coulomb force the electrostatic inverse-square interaction between electrically charged particles, discovered in 1785 by the French physicist Charles Augustin de Coulomb

coupling constant the strength of an interaction between particles or fields; electric charge and Newton's gravitational constant are coupling constants

coupling strength the strength of the force exerted in an interaction between particles

cross-section in a collider, the area of a material with nuclei that acts as a target for a beam of particles hitting it

dark energy a mysterious form of energy that has been associated with negative pressure vacuum energy and Einstein's cosmological constant; it is hypothesized to explain the data on the accelerating expansion of the universe; according to the standard model of cosmology, the dark energy, which is spread uniformly throughout the universe, makes up about 70 percent of the total mass and energy content of the universe

dark matter invisible, not-yet-detected unknown particles of matter, representing about 25 percent of the total matter-energy in the universe according to the standard model; its presence is necessary if Newton's and Einstein's gravity theories are to fit data from galaxies, clusters of galaxies, and cosmology that show much stronger gravity than is predicted by the theories; together, dark matter and dark energy mean that 96 percent of the matter and energy in the universe is invisible

degree of freedom an independent parameter in a physical system that describes that system's configuration

Dirac equation a relativistic wave equation formulated by Paul Dirac in 1928 that is consistent with both quantum mechanics and the theory of special relativity; it describes the quantum physics of elementary spin-½ particles such as electrons, and can be generalized to apply to the curved spacetime of Einstein's general relativity; the wave equation also predicted the existence of antimatter, such as the positron

drift chamber (wire chamber) a proportional counter with a wire under high voltage running through a metal conductor enclosure that is filled with a gas (such as an argon–methane mix), so that an ionizing particle will ionize surrounding atoms; the ions and electrons are accelerated by an electric field acting on the wire, producing an electric current proportional to the energy of the particle that is to be detected, which allows an experimentalist to count particles and determine their energy

electromagnetism one of the four fundamental forces of nature, the others being gravitation, the weak interaction (radioactivity), and the strong interaction; the electromagnetic field interacts through the photon, which is its force carrier, with the electric charge of elementary particles such as the electron

electroweak theory the unified description of the electromagnetic and weak interactions of particles; although they appear to be different forces at low energies, they merge into one above the unification energy 246 GeV

elementary particle a particle that does not have any substructure such as smaller particles; considered one of the basic building blocks of matter in the universe

energy cutoff a maximum value of energy, momentum, or length in particle physics, usually chosen to make infinite quantities in calculations become finite

ether (or aether) the medium through which it was believed for centuries that energy and matter moved, something more than a vacuum and less than air; its origins were in the Greek concept of "quintessence," but during the late 19th century, the Michelson–Morley experiment disproved the existence of the ether

femtobarn   a unit of cross-section area where a barn is equal to 10^{-28} m², a millibarn (mb) is equal to 10^{-31} m², and a femtobarn (fb) is equal to 10^{-43} m²

fermion   a particle with half-integer spin like protons and electrons, which make up matter

field   a physical term describing the forces between massive bodies in gravity and electric charges in electromagnetism; Michael Faraday discovered the concept of field when studying magnetic conductors

fine-tuning   the unnatural cancelation of two or more large numbers involving an absurd number of decimal places, when one is attempting to explain a physical phenomenon; this signals that a true understanding of the physical phenomenon has not been achieved

gauge boson   in the standard model of particle physics, a carrier of a force, such as the photon, which is the carrier of the electromagnetic interaction; the W and the Z bosons, which are the carriers of the weak interaction; and the gluon, which is the carrier of the strong interaction

gauge invariance   in electromagnetism, the property that a class of scalar and vector potentials, related by gauge transformations, retains the same electric and magnetic fields; Maxwell's field equations are such that the electric field and the magnetic field can be expressed in terms of the derivative of a scalar field (scalar potential) and a vector field (vector potential); gauge invariance has been extended to more general theories such as nonabelian Yang–Mills fields and gravitational fields

Geiger counter   a particle detector that measures ionizing radiation, such as the emission of alpha particles, beta particles, or gamma rays by atomic nuclei

general relativity   Einstein's revolutionary gravity theory, created in 1916 from a mathematical generalization of his theory of special relativity; it changed our concept of gravity from Newton's universal force to the warping of the geometry of spacetime in the presence of matter and energy

GeV   a unit of energy equal to one billion electron volts, or a gigaelectron volt

gluon   the exchange particle or gauge boson for the strong force that binds quarks to make hadrons; gluons are analogous to photons, which carry the electromagnetic force between two electrically charged particles

graviton   the hypothetical smallest packet of gravitational energy, comparable to the photon for electromagnetic energy; the graviton has not yet been seen experimentally

gravity   as first expressed by Isaac Newton, a force by which physical bodies attract each other proportional to their masses, and inversely proportional to the square of the distance between them; Einstein's general relativity theory described gravitation as the curvature of spacetime by matter; gravity is the weakest of the known four fundamental forces

group (in mathematics)   in abstract algebra, a set that obeys a binary operation that satisfies certain axioms; for example, the property of addition of integers makes a group; the branch of mathematics that studies groups is called *group theory*

hadron   a particle composed of quarks bound together by the strong force; there are two kinds: baryons, such as protons and neutrons, made of three quarks, and mesons, such as pions, made from one quark and one antiquark; in Greek, hadrós means "stout" or "thick"

Higgs field the field associated with the Higgs boson, which is theorized to impart mass to the known elementary particles of the standard model

Higgs particle or boson theorized to be an elementary particle by Peter Higgs in 1964, its probable discovery was announced on July 4, 2012, and confirmed more confidently on March 14, 2013; this particle plays a pivotal role in the standard model

horizon problem identified in the standard Big Bang model of cosmology during the late 1960s, it occurs because widely separated regions of the early universe had the same temperature and other physical properties, yet were unable to communicate with each other because of the large distances between them; the finite measured speed of light prevents the regions from being connected causally; inflation theory and the variable speed of light theory have been proposed to resolve the horizon problem

identity in mathematics, an equality relation A = B, such that A and B contain numbers or variables; A = B is an identity when it is true for all values of the functions making up A and B; in algebra, the identity element e when combined with any element x of S gives the same x: ex = xe = x for all x in S

inflation a theory proposed during the early 1980s by Alan Guth and others to resolve the flatness, horizon, and homogeneity problems in the standard Big Bang model; the very early universe is pictured as expanding exponentially fast in a fraction of a second

inverse femtobarn (fb^{-1}) a measurement of particle collision events per femtobarn of target cross-section, which serves as the conventional unit for time-integrated luminosity

isotopic spin or isospin proposed by Werner Heisenberg in 1932 to explain the properties of the newly discovered neutron, it is a quantum number related to the strong interaction such that particles with different charges that are affected equally by the strong force can be treated as different states of the same particle, with isospin values determined by the number of charge states; physically, isospin is a dimensionless quantity—in other words, it does not have the units of angular momentum or spin, but its name comes from the fact that the mathematics of isospin are similar to those describing spin

kinetic energy the energy a particle has during its motion; defined as the work needed to accelerate a particle with mass from rest to its stated velocity.

K meson (K* meson) an elementary particle in the class of pseudoscalar mesons (K spin 0) or pseudoscalar vector mesons (K* spin 1) distinguished by the strangeness quantum number; it is a bound state of a strange quark (or antiquark) and an up or down quark (or antiquark)

Lagrangian the function that describes the dynamics of a physical system; it is named after French mathematician Joseph Louis Lagrange, who reformulated classical mechanics

large electron positron (LEP) collider a multinational, circular collider at CERN, built in a tunnel straddling the border of Switzerland and France; operating from 1989 to 2000, it was the largest accelerator of leptons ever built, reaching an energy of 209 GeV by 2000

large hadron collider (LHC) the world's largest high-energy accelerator, built at CERN from 1998 to 2008 using the former LEP circular tunnel; built by a collaboration of more than 10,000 scientists and engineers from about 100 countries, it has a circumference of 27 km (17 mi) beneath the French–Swiss border near Geneva

lepton an elementary particle with spin ½ that does not undergo strong interactions and satisfies the Pauli exclusion principle—in other words, no two leptons can occupy the same quantum state; the electron, muon, and tau are charged leptons whereas the neutrinos are electrically neutral; the electron governs the chemical properties of atoms; there are six types or flavors of leptons, forming three generations or families, matching the three generations of quarks

local (locality) the concept that a particle's position and momentum can be localized at a point in space; standard quantum field theory is based on the axiom of locality, in which two particles separated by a distance in spacetime cannot interact; this is in accordance with special relativity, in which no particle can exceed the speed of light; for example, the electromagnetic force between charged particles is mediated by photons with a finite speed, which makes the force local, in contrast to Newton's formulation of gravity, in which the force of attraction between massive bodies acted instantaneously (non-locally)

Lorentz invariance invariance of physical laws under Lorentz transformations from one inertial frame to another

Lorentz transformations mathematical transformations from one inertial frame moving with uniform velocity to another inertial frame such that the laws of physics remain the same; named after Hendrik Lorentz, who developed them in 1904, these transformations form the basic mathematical equations underlying special relativity

luminosity in astronomy, the total amount of energy emitted by an astronomical object per unit time; in particle accelerators, the number of collisions per unit area of the target cross-sections over time; the number of collisions can be calculated by multiplying the integration over time of the luminosity by the total cross-section for the collisions; at the LHC, this count is measured as the inverse femtobarns for the time period

mass scale the mass or energy scale in particle physics at which the strengths of particle forces take on characteristic properties; for example, the Planck energy (mass) of 1.22×10^{19} GeV (Planck mass, 2.18×10^{-8} kg) is the energy when the quantum effects of gravity are expected to become strong

matrix in mathematics, a rectangular array of symbols, such as numbers, arranged in rows and columns.

meson a short-lived boson composed of a quark and an antiquark, believed to bind protons and neutrons together in the atomic nucleus

Michelson–Morley experiment an experiment conducted in 1887 by Albert Michelson and Edward Morley that proved that the ether did not exist; beams of light traveling in the same direction, and in perpendicular directions, in the supposed ether showed no difference in speed or arrival time at their destination

Minkowski light cone the path that light, emanating from a single event E, would take through four-dimensional spacetime (three space dimensions and one time dimension); confined to a two-dimensional plane, the light spreads out in a circle from E;

if we graph the growing circle with the vertical axis representing time and the horizontal axis representing space, the spreading light forms two cones, one of which is the past light cone and the other the future light cone; the sides of the cones represent light traveling at speed c, and massive particles cannot pass through them for, in special relativity, no particle can travel faster than the speed of light

MOG a relativistic modified theory of gravitation that generalizes Einstein's general relativity and can fit astronomical and cosmological data without exotic dark matter; MOG stands for "modified gravity"

multiverse a hypothetical set of finite or infinite universes including the universe we inhabit; currently it is postulated as a paradigm that allows for fine-tuning of the calculation of parameters and constants in particle physics, and is closely related to the anthropic principle

muon from the Greek letter mu (μ), an unstable elementary particle with spin-½ similar to the electron; it belongs to the family of leptons, together with the neutrinos and the tau

neutralino a hypothetical particle predicted to exist by supersymmetry that is a popular dark-matter particle candidate; there are four neutralinos that are electrically neutral fermions, the lightest of which is stable

neutrino an elementary particle with zero electric charge and a tiny mass that has not yet been measured accurately; very difficult to detect, it is created in radioactive decays and is able to pass through matter almost without disturbing it; there are three flavors of neutrinos: ν_e, ν_μ, and ν_τ

neutron an electrically neutral particle found in the atomic nucleus, and having about the same mass as the proton

Noether's theorem a mathematical theorem stating that any symmetry of the action of a physical system (the integral over the Lagrangian density function) has an associated conservation law; for example, the rotational invariance of the action leads to the conservation of angular momentum; the theorem was published in 1918 by the German mathematician Emmy Noether

nonlocality the direct action-at-a-distance between two particles that are separated in space with no mediating mechanism; Newton's formulation of gravity was nonlocal, as the force of attraction between massive bodies acted instantaneously; quantum mechanics, too, acts nonlocally, through the quantum entanglement of two photons or electrons, which Einstein called "spooky action at a distance"; nonlocal quantum field theory can be formulated consistently as a finite theory and may violate causality at small distances

parity a symmetry property of particles under spatial inversion

perturbation theory a mathematical method for finding an approximate solution to an equation that cannot be solved exactly, by expanding the solution in a series in which each successive term is smaller than the preceding one

phase of waves in sinusoidal shapes of waves, the initial angle of a sinusoidal function at its origin is termed the *phase difference*, which is the fraction of a wave cycle that has elapsed relative to the origin

photon the quantum particle that carries the energy of electromagnetic waves; the spin of the photon is 1 times Planck's constant h.

pi meson (pion)   the lightest, unstable spin-0 pseudoscalar (negative parity) meson; the positively charged pion is composed of an up quark and an antidown quark

Planck length   a constant unit of length equal to $1.616199(97) \times 10^{-35}$ m first described by Max Planck; it is defined using the three fundamental constants: the gravitational constant G, the speed of light c in a vacuum, and Planck's constant h

Planck mass or energy   the Planck constant formed from G, c, and h, expressed as a unit of mass or energy

Planck's constant (h)   a fundamental constant that plays a crucial role in quantum mechanics, determining the size of quantum packages of energy such as the photon; it is named after Max Planck, a founder of quantum mechanics; the symbol ħ, which equals h divided by 2π, is used in quantum mechanical calculations

positron   the antiparticle of the electron with positive electrical charge and spin ½; it is a stable particle with the same mass as the electron

potential energy   in gravitation, the stored energy of a particle when it is held at an elevated position; other force fields such as electromagnetism also have potential energy

proton   a particle that carries a positive electric charge and is the nucleus of a hydrogen atom; it is a spin-½ hadron composed of two up quarks and one down quark.

pseudoscalar   the property of a particle that changes sign under a spatial parity inversion; this is in contrast to the scalar property of a particle that does not change sign

quantum chromodynamics (QCD)   a theory of the strong-interaction color force, it describes the interactions between quarks and gluons that make up hadrons; QCD is described by a nonabelian SU(3) Yang–Mills gauge theory, which consists of colored gluon fields and confined quarks

quantum electrodynamics (QED)   the relativistic quantum field theory of the electromagnetic field, describing how charged particles interact with photons at the quantum level; QED makes extremely accurate predictions, such as the anomalous magnetic moment of the electron and the Lamb shift of the energy levels of hydrogen

quantum field theory   the modern relativistic version of quantum mechanics used to describe the physics of elementary particles; it can also be used in nonrelativistic field-like systems in condensed-matter physics

quantum mechanics   the theory of the interaction between quanta (radiation) and matter; the effects of quantum mechanics become observable at the submicroscopic distance scales of atomic and particle physics, but macroscopic quantum effects can also be seen in the phenomenon of quantum entanglement

quantum spin   the intrinsic quantum angular momentum of an elementary particle; this is in contrast to the classical orbital angular momentum of a body rotating about a point in space.

quark   the fundamental constituent of all particles that interact through the strong nuclear force; quarks have spin-½, are fractionally charged, and come in several varieties or "flavors" called up, down, charm, strange, top, and bottom; the lightest quark is the up quark; because of their color charge, they are confined within particles such as protons and neutrons, so they cannot be detected as free particles; quarks, like leptons, form three generations

quark–gluon plasma   a phase of quantum chromodynamics that exists at an extremely high temperature and density, such as at the beginning of the universe

renormalization   in quantum field theory, a technique used to deal with infinities arising in calculated quantities; it was first developed in quantum electrodynamics to treat infinities occurring in the calculation of the charge and mass of an electron

rotational invariance   the property of a physical system when it behaves the same regardless of how it is oriented in space; if the action (the integral over time of the Lagrangian) of a system is invariant under rotations, then by Noether's theorem, the angular momentum is conserved

scalar field   a physical term that associates a value without direction to every point in space, such as temperature, density, and pressure; it is in contrast to a vector field, which has a direction in space; in Newtonian physics or in electrostatics, the potential energy is a scalar field and its gradient is the vector force field; in quantum field theory, a scalar field describes a boson particle with spin zero; see *vector field*

S-matrix   connecting the initial and final states of a scattering process, it is defined technically as the unitary matrix, relating asymptotic particle states in the Hilbert space of physical states

special relativity   Einstein's initial theory of relativity, published in 1905, in which he explored the "special" case of transforming the laws of physics from one uniformly moving (inertial) frame of reference to another; he called it "special" because it did not include gravity, or accelerated frames of reference; the equations of special relativity revealed that the speed of light is a constant, that objects appear contracted in the direction of motion when moving at close to the speed of light, and that $E = mc^2$, or energy is equal to mass times the speed of light squared

spin   see *quantum spin*

spontaneous symmetry breaking   the breaking of a symmetric state into an asymmetric state; for example, the equations of motion or the Lagrangian of a physical system obey rotational symmetry, but the lowest energy or ground-state solutions of the equations of motion do not have that symmetry; spontaneous symmetry breaking plays a pivotal role in the electroweak theory in the standard model

standard deviation   represented by the symbol sigma (σ), in statistics and probability theory it tells how much variation or dispersion exists from the mean or average value of an experimental result

standard model of particle physics   a model of the electromagnetic, weak, and strong interactions based on the dynamics of the known quarks, leptons, and gauge bosons, as well as the Higgs boson; developed by a collaborative effort by many particle theorists and experimentalists throughout the mid to late 20th century, it has been very successful in explaining a wide variety of experiments

string theory   a theory based on the idea that the smallest units of matter are not point particles, but vibrating strings; a popular research pursuit in physics for several decades, string theory has some attractive mathematical features, but has yet to make a testable prediction

strong force   one of the four fundamental forces, the strong interaction (or nuclear force) is about 100 times stronger than the electromagnetic force and orders of magnitude stronger than the weak and gravitational forces; at the nuclear level, it binds protons and neutrons together to form the nucleus of an atom; at a smaller scale, it is the force mediated by colored gluons that binds quarks together to form hadrons such as protons and neutrons

superconductivity the physical phenomenon of zero electrical resistance and the expulsion of magnetic fields in certain materials cooled below a critical temperature; a quantum mechanical phenomenon discovered by Dutch physicist Heike Kamerlingh Onnes in 1911, its theoretical explanation was proposed in 1957 by Bardeen, Cooper, and Schrieffer

supergravity a field theory that combines gravity (general relativity) and supersymmetry; it contains a spin-2 field with a quantum particle, the graviton, which supersymmetry demands has a superpartner with spin ³/₂ called the *gravitino*

superstring theory an attempt to unify the four fundamental forces of nature in a version of string theory that incorporates fermions as well as bosons in a supersymmetric framework; in it, particles are modeled as tiny vibrating strings

supersymmetry a theory developed during the 1970s that, proponents claim, describes the most fundamental spacetime symmetry of particle physics: for every boson particle there is a supersymmetric fermion partner, and for every fermion there exists a supersymmetric boson partner; to date, no supersymmetric particle partner has been detected

symmetry breaking in nature, the breaking of a symmetry of a physical system; there are two kinds of symmetry breaking: explicit, in which the physical laws of a system are not invariant under a symmetry generated by a set of transformations, and spontaneous symmetry breaking, in which the physical laws *are* invariant under a given set of transformations, but the vacuum state or lowest energy state (ground state) is not invariant

symmetry group mathematically, the group of all isometries (the congruence of two geometrical figures) under which an object is invariant

synchrotron a type of particle accelerator developed from the cyclotron, in which the guiding magnetic field, which bends the particles into a closed path, is time dependent, and is synchronized to a particle beam with increasing kinetic energy; it is often the initial accelerator in high-energy colliders

TeV a measure of energy equal to one trillion electron volts or $1.60217657 \times 10^{-7}$ joules (J), called a teraelectron volt

Tevatron accelerator a circular particle accelerator at the Fermi National Accelerator Laboratory (Fermi Lab) in Batavia, Illinois, that was the second-largest high-energy accelerator in the world after the large hadron collider at CERN; a synchrotron that accelerated protons and antiprotons in a 6.86-km (4.26-mi) ring to beam energies of up to about 1 TeV (which explains its name), it was completed in 1983 and closed down in September 2011

unitarity a restriction in particle physics on the permitted evolution of a quantum system that guarantees that the sum of all probabilities of a physical process (such as a scattering process) always equals unity, or one; the S-matrix describing a scattering of particles must be a unitary operator

vacuum in quantum mechanics, the lowest energy state, which corresponds to the vacuum state of particle physics; in modern quantum field theory, it is the state of perfect balance of the creation and annihilation of particles and antiparticles

vacuum expectation value   in quantum field theory, the average quantum value of a field operator in a vacuum

variable speed of light (VSL) cosmology   an alternative to inflation theory in which the speed of light was much faster at the beginning of the universe than it is today; like inflation, this theory solves the horizon and flatness problems in the very early universe in the standard Big Bang model

vector field   a field with direction in space, such as the force field of gravity or the electric and magnetic force fields in Maxwell's field equations; see *scalar field*

void cosmology   an alternative to the standard LambdaCDM cosmology, in which we inhabit a large cosmic void that is surrounded by matter in the form of galaxies; in a simple version of the model, the void is a spherically symmetric "bubble" described by the exact Lemaître–Tolman–Bondi cosmological solution of Einstein's field equations; agreeing well with available cosmological data, the model does not hypothesize the existence of dark energy or an accelerated expansion of the universe

W and Z bosons   the intermediate vector (spin-1) bosons of electroweak theory; the W boson is electrically charged (W^+, W^-) and was discovered together with the electrically neutral Z boson at CERN in 1983; both the W and Z bosons are very short-lived, with a half-life of about 3×10^{-25} seconds

Ward identity   a mathematical identity in quantum field theory that follows from gauge invariance and is valid after renormalization of the theory; also known as the Ward–Takahachi identity, developed by John Ward and Yasushi Takahashi to guarantee the cancelation of ultraviolet divergences in all orders of perturbation theory

weak interaction or weak force   one of the four fundamental forces of nature, it is responsible for the radioactive decay of nuclei and nuclear fusion; the weak and electromagnetic interactions have been united in the electroweak theory of the standard model

FURTHER READING

Aczel, Amir D. *Present at the Creation: The Story of CERN and the Large Hadron Collider.* New York: Crown Publishers, 2010.

Baggott, Jim. *The Quantum Story: A History in 40 Moments.* New York: Oxford University Press, 2011.

Baggott, Jim. *Higgs: The Invention & Discovery of the "God Particle."* Oxford: Oxford University Press, 2012.

Carroll, Sean. *The Particle at the End of the Universe: How the Hunt for the Higgs Boson Leads Us to the Edge of a New World.* New York: Dutton (Penguin Group), 2012.

Close, Frank. *Particle Physics: A Very Short Introduction.* New York: Oxford University Press, 2004.

Close, Frank. *The New Cosmic Onion: Quarks and the Nature of the Universe.* New York: Taylor & Francis, 2007.

Close, Frank. *The Infinity Puzzle: How the Hunt to Understand the Universe Led to Extraordinary Science, High Politics, and the Large Hadron Collider.* New York: Basic Books, 2011.

Coughlan, Guy D., James E. Dodd, and Ben M. Gripaios. *The Ideas of Particle Physics: An Introduction for Scientists,* 3rd ed. New York: Cambridge University Press, 2006.

Cox, Brian, and Jeff Forshaw. *The Quantum Universe (and Why Anything That Can Happen, Does).* Boston, MA: Da Capo Press (Perseus Books Group), 2011.

Crease, Robert P., and Charles C. Mann. *The Second Creation: Makers of the Revolution in Twentieth-Century Physics,* rev. ed. New Brunswick, NJ: Rutgers University Press, 1996 (originally published 1986).

Fayer, Michael D. *Absolutely Small: How Quantum Theory Explains Our Everyday World.* New York: AMACOM, 2010.

Fritzsch, Harald. *Elementary Particles: Building Blocks of Matter.* Singapore: World Scientific Publishing, 2005.

Johnson, George. *Strange Beauty: Murray Gell-Mann and the Revolution in Twentieth-Century Physics.* New York: Alfred A. Knopf, 1999.

Krauss, Lawrence M. *Quantum Man: Richard Feynman's Life in Science.* New York: W.W. Norton, 2012.

Kumar, Manjit. *Quantum: Einstein, Bohr, and the Great Debate about the Nature of Reality*. New York: W.W. Norton, 2008.

Lederman, Leon, with Dick Teresi. *The God Particle: If the Universe Is the Answer, What Is the Question?* Boston, MA: Houghton Mifflin, 2006.

Lincoln, Don. *The Quantum Frontier: The Large Hadron Collider*. Baltimore, MD: Johns Hopkins University Press, 2009.

Pickering, Andrew. *Constructing Quarks: A Sociological History of Particle Physics*. Chicago, IL: The University of Chicago Press, 1984.

Randall, Lisa. *Higgs Discovery: The Power of Empty Space*. New York: HarperCollins, 2013.

Sample, Ian. *Massive: The Hunt for the God Particle*. Virgin Digital, 2010.

Schumm, Bruce A. *Deep Down Things: The Breathtaking Beauty of Particle Physics*. Baltimore, MD: Johns Hopkins University Press, 2004.

Veltman, Martinus. *Facts and Mysteries in Elementary Particle Physics*. Singapore: World Scientific Publishing, 2003.

Weinberg, Steven. *Dreams of a Final Theory: The Search for the Fundamental Laws of Nature*. New York: Pantheon Books, 1992.

Wouk, Herman. *A Hole in Texas*. New York: Little, Brown, 2004.